Wave-Swept Shore

The publisher gratefully acknowledges

the generous contribution to this book

provided by the Gordon and Betty Moore

Fund in Environmental Studies

Wave-Swept Shore

THE RIGORS OF LIFE ON A ROCKY COAST

for Matt,
enjoy the rocky coast!
Anne W. Rosenfeld
3·2006

Text by MIMI KOEHL

Photographs by ANNE WERTHEIM ROSENFELD

University of California Press BERKELEY LOS ANGELES LONDON

University of California Press, one of the most distinguished university presses in the United States, enriches lives around the world by advancing scholarship in the humanities, social sciences, and natural sciences. Its activities are supported by the UC Press Foundation and by philanthropic contributions from individuals and institutions. For more information, visit www.ucpress.edu.

University of California Press
Berkeley and Los Angeles, California

University of California Press, Ltd.
London, England

Library of Congress Cataloging-in-Publication Data

Koehl, Mimi, 1948–.
    Wave-swept shore : the rigors of life on a rocky coast / text by Mimi Koehl ; photographs by Anne Wertheim Rosenfeld.
        p.   cm.
    Includes bibliographical references (p.   ) and index.
    ISBN 0-520-23812-5 (cloth : alk. paper).
    1. Intertidal ecology. I. Rosenfeld, Anne Wertheim, 1951–. II. Title.
QH541.5.S35K67   2006
577.69'9—dc22   2005002456

Manufactured in China

14   13   12   11   10   09   08   07   06
10   9   8   7   6   5   4   3   2   1

The paper used in this publication meets the minimum requirements of ANSI/NISO z39.48–1992 (R1997) (Permanence of Paper).

Contents

Acknowledgments

We thank Robert Paine for introducing us to each other and for so generously teaching us over the years (on many a wild, wet field trip) about the ecology and natural history of exposed rocky shores.

I thank Ellen Daniell, Harry Greene, Mark Harrington, Mark Moffett, Rachel Norton, Zack Powell, Alan Shanks, and Richard Strathmann for plowing through early drafts of this book and making many helpful suggestions. I am also grateful to my mentors in biomechanics, Stephen Wainwright, Steven Vogel, and John Currey, who started me along the research path that eventually led to this book. I appreciate the many field assistants who have helped me schlep electronic gear over slippery rocks down to the waves. Thanks to Scott Jackson for technical assistance and good humor in the face of deadline panics. Much of my research reported here was supported by grants from the National Science Foundation, by a Guggenheim Fellowship, and by a MacArthur Fellowship.

Mimi Koehl

Photography is, among other things, a form of visual scholarship. Constructing a book, and continually looking at the relationships of its pliant, evolving parts, is a building process. For encouraging both the builder and artist in me, I continue to hold a special place in my heart for the late Ann O'Hanlon.

I would like to thank everyone who helped me manage the abundant challenge of getting photographic gear to and around the intertidal site described in this book. In particular, I would like to thank Ryan Baldwin, without whom it would have been impossible to carry this project forward. I deeply appreciate his willingness to work any hour of the day or night in any sort of wet, windy, slippery, or otherwise trying conditions. I'm also grateful for his indispensable help carrying huge amounts of equipment—uncomplainingly—for miles at a stretch. Most important, Ryan's friendship, his extensive knowledge of photography, and his technical and aesthetic assistance have been an unfailing support throughout this project.

I would like to thank Laura Shapiro, who is consistently a most sensible and helpful guide in every direction of my work and the dearest of friends. Her contributions to this and other projects have been innumerable.

Finally, I would like to thank my husband, Bob Rosenfeld, and our son, Matthew, for their contributions to this book, many but not all of them age-related. A tide-driven work schedule makes for numerous domestic disruptions, and their tolerance has been admirable.

Anne Rosenfeld

Wave-Swept Shore

1

PLACE AND SCALE

Getting Down to the Shore

We are creatures of the land. We view the ocean as a vast expanse of water stretching to the horizon at the edge of our world. We've been taught that the ocean is teeming with life, but for most of us that marine menagerie is hidden from view, except at the places where water meets land. At this junction of the terrestrial and aquatic realms, we find a fascinating collection of animals and plants[1] that face the challenges of both worlds: they are exposed to the rigors of wind, sun, and rain when the waters recede at low tide, but they also must withstand the blows of crashing waves when the sea rises over them again at high tide. The strip of shore between

---

1. In marine habitats we find a variety of vascular plants, such as the surfgrasses that are related to the grasses we see in lawns and meadows on land. Plants like grasses, trees, and cabbages are called *vascular plants* because they have distinctive plumbing systems that conduct water through their bodies. On a rocky seashore we also find many types of algae, which look like plants but which don't have such plumbing systems and which are actually members of a different kingdom. Like vascular plants, the algae make their living by photosynthesis, trapping energy from the sun. For ease of discussion in this book, we often use the generic term *plant* to refer to both the vascular grasses and the algae on the shore. Sometimes we use *seaweeds* as a synonym for *algae,* and we mention *kelp,* which are members of a particular group of large brown algae.

the high- and low-tide marks, where plants and animals spend part of their time submerged in water and part of it exposed to air, is called the *intertidal zone*. This brutal habitat, where the dynamics of the physical environment are so powerful and so obvious, is an exciting place to explore how living things interact with their surroundings.

On the scale of living things, humans are large. Our intuitions about the physical world are based on how we experience it, but these intuitions can mislead us when we try to understand how smaller creatures encounter their environment. We stand on the shore at low tide, squinting at the sun as the wind whips through our hair, while a snail hunkered down in a crack in the rock at our feet is hiding in an environment that is shady and still. A rogue wave knocks us off our feet, but it skims over the crack and leaves the snail unperturbed. Side by side on the same chunk of shore, the little snail shelters in a very different physical environment from the one that we bigger humans experience.

A *microhabitat* is the neighborhood of a living thing: the local terrain, the plants and animals living next door, and the physical conditions in the immediate vicinity of the creature. What is it like in the neighborhood of a seaweed, a barnacle, or a starfish on a rocky shore? When the tide is out, how fast is the wind, how hot and humid is the air, and how bright is the sunlight encountered by each of these living things? When the tide is in, how hard are they hit by the waves, and how much light penetrates through the water above them?

In this book we explore the wealth of diverse microhabitats that can be found on just one small section of the northern California shoreline, a stretch of coast less than a kilometer in length. By examining the animals and plants surviving in these microhabitats, we can explore some of the fundamental ways in which living things interact with the physical world

2 The edge of the sea *(opposite)*.

3 The rough topography of a rocky shore is revealed at low tide. Picking your way across this tricky terrain, you may crunch through strips of shells and sand, which in this picture look like pale horizontal streaks between the boulders and the cliffs. Clambering across a field of boulders, you teeter on the loose rubble that wobbles under your feet and slither across the larger rocks carpeted with slippery seaweeds. You may skirt around big islands of stone like the one jutting into the right side of the photograph. This massive rock bench has wide horizontal surfaces and sheer vertical faces, some baking in the sun and others hidden in the shadows. Fine cracks and gaping crevasses cut across the stony landscape, and tide pools large and small pock the surfaces. Watery channels, like those shimmering in the foreground, wend their way between the rocks.

around them. By showing you the richness and complexity of one special place that captivates us, we hope to provide you with new ways to experience and appreciate the rich tapestry of microhabitats and living things in places you can explore for yourself. The way we look for different microhabitats at a particular coastal site and the way we figure out what the environment is like within each one are ways of looking at nature that will work just as well in other places, from your own backyard or neighborhood park to the alpine meadows, shady forests, crystal lakes, and wave-swept beaches you might visit on vacation.

Leave the path across the hills for a while and climb down to the edge of the sea at low tide. Dip your hand into pools and feel the warmth or chill of the water. Run your fingers through a damp clump of algae; then lay your palm on a sun-baked field of barnacles. Stick your nose into a sea cave and breathe the cool, fishy mist. Gather wet drips on your sleeve in the shade of an overhanging rock. Feel the rubbery stretchiness of kelp, the calcified stiffness of mussel shells, and the fleshy softness of sea anemones. Watch as the tide comes in and everything changes. As you retreat from the rising waters, listen to the thud and swash of waves hurling themselves onto the shore. Feel the sea spray chilling your skin. Trace the path of swirling bubbles as water rushes through channels and tarries behind rocks. Try to feel the environment met by the animals and plants clinging to these rocks—living things so different from us.

Looking Closer

Look where the land and sea meet along a stretch of rocky, wave-swept coastline. The topography of the shore is complex. Boulders of all sizes are strewn on the sand. Broad benches and vertical walls of rock are interspersed with

4 Rough or calm, water or air, sun or shade: conditions vary within just one small stretch of coastline.

gently sloping stone faces. Some areas glisten in bright sun, while others are in shadow; some are underwater, while others are exposed to air; some are pounded by waves, while others are protected from roiling currents (plate 4).

Let's scramble down for a closer look at one rock bench near the water's edge at low tide (plate 5). Clusters of algae, mussels, and goose barnacles, with their armor of white plates, carpet the exposed surfaces of the bench. Starfish huddle in crevices in the rock, while sea anemones hide in tide pools sprinkled across the bench. Mussels and goose barnacles are tightly packed together (plate 6). The upper surfaces of their shells are chalky dry in the sun, while the spaces between them are silty and damp. Other creatures make their home in the complex architecture of this clump of mussels: acorn barnacles encrust the mussel shells, while worms, crustaceans, and snails shelter in the recesses of the mussel bed. If you peek into a shady crevice cut into the rock, you see that this dark recess harbors a very different microcosm (plate 7). In this cool, damp refuge from waves and wind, a pink patchwork of delicate sponges and algae thrives. If you lean over one of the tide pools in the same rock bench, you discover a water-filled refuge where sea anemones spread their crowns of tentacles between emerald strands of surfgrass and rose-colored bushes of coralline algae (plate 8). If you watch the pool patiently, you may glimpse small animals moving among the attached anemones and plants—perhaps a colorful sea slug gliding gracefully (plate 9), a tiny crab dashing across the bottom, or a little fish flitting through the water. Thus what seems to be a simple rock bench at the edge of the sea holds an array of different microhabitats.

Now climb above the rock bench to inspect the sunny face of another rock, higher up in the intertidal zone. Barnacles pave the stone between turfy clumps of algae (plate 11). A closer look reveals crevices where limpets congregate at low tide (plate 12).

5 A wealth of microhabitats, each characterized by different physical conditions, occurs on a single rock bench, seen here at low tide. For example, the microhabitat in the crevice where the orange starfish hides is shady, cool, and damp when the tide is out, and it is protected from the brunt of the waves that pound this bench when the tide comes in. In contrast, the white goose barnacles (barnacles with fleshy stalks) and black mussels cover areas of the rock exposed to crashing waves at high tide and howling wind at low tide, when they bake in the bright sun. However, a very different microhabitat exists deep within the interstices of the mussel bed, where small animals lurk in dark, damp pockets, protected from sun, wind, and waves *(above)*.

6 Goose barnacles, each protected by a tilework of white plates, and black-shelled mussels are densely squeezed together as they compete for limited space on the rock. This mat of animals provides a complex three-dimensional structure in which other creatures make their home *(opposite)*.

7 The wall of a crevice that remains damp and shady at low tide harbors purple sponges and a variety of algae, from slippery dark fronds and calci-fied pink bushes to crusts that look like tar smeared on the rock *(above)*.

8 In the water of a tide pool, sea anemones inflate their crowns of green tentacles among shrubby pink coralline algae and bright seagrass blades, while shiny seaweed fronds above the pool dry in the air. Coralline algae are species of red algae that have calcium crystals stiffening their tissues, so they look chalky and pink rather than fleshy and dark like other species of algae that are not calcified *(opposite)*.

9 A delicate sea slug cruises along the pebbles at the bottom of a tide pool *(above)*.

10 A rock face, higher above the water than the bench in plate 5, cooks in the sun at low tide *(opposite)*.

11 Clusters of small white acorn barnacles cover the sun-baked rock
between clumps of algae. Fronds of one species of algae hang like wilted
bouquets from the crevices, while dense patches of another species look like
pieces of turf cut from a lawn *(opposite)*.

12 Limpets (snails with cone-shaped shells) cluster together by a shady, moist
crack in the face of a sunny rock *(above)*.

13 Overview, at low tide, of a large rock facing the ocean. Some water has been trapped in the tide pool at lower left. Zones of various colors and textures are visible at different heights above the water. Each zone is populated by a distinct assemblage of plants and animals. The vertical black rift cutting across these zones is the mouth of a cave.

When the tide has ebbed to its lowest level, thread your way carefully across slippery seaweeds and sharp-edged mussels back down to the water's edge. Stand back and look at a massive vertical rock facing the ocean. As you scan its face from bottom to top, you notice distinct zones of different colors and textures. Low in the intertidal zone, crusts of pink coralline algae cover the rock under canopies of slick, dark seaweed fronds and in shimmering pools full of apple-green sea anemones. Higher in the intertidal, clusters of mussels and goose barnacles form a bumpy black and white necklace around the rock. Higher still, a band of acorn barnacles forms another zone that encircles the rock like a bathtub ring. The shadowy mouth of a sea cave cuts through these horizontal stripes of animals and plants on the face of the rock.

Three steps into the cave, you find yourself in yet another world. The zones of creatures you saw just outside its mouth are gone. Once your eyes adjust to this misty, dark hollow, you see that the dank recesses of the cave are inhabited by a stunning array of living things (plate 14). Although dim light filters in through the mouth behind you, through a hole opening to the sky above you, and through a small window ahead of you at the far end of the cave, no seaweed turfs or kelp beds grow here. The only algae in this twilight world are the magenta crusts (coralline algae) and tar-like splashes (fleshy red algae) that paint the rocks. Now carefully pick your way through this colorful cavern, trying not to step on the slippery patches of fragile plants and animals living on its surfaces. At the far end of the cave, crane your neck to take a close look at the ceiling just above the narrow, nozzle-like window exiting the cave. Here, squat acorn barnacles, flat limpets, and stumpy clumps of goose barnacles hug the rock (plate 15). Now turn around and contrast the crusty citizens of the ceiling with the inhabitants of the cave's wide belly (plate 16). The damp walls are carpeted

14 As we stand with our backs to the ocean in the mouth of the cave shown in plate 13, we can peer into its dark, damp recesses. At the opposite end of the cave, light filters in through a small, windowlike opening. When the tide is high, water flows in the mouth where we stand and across the wide belly of the cave, and then rushes out through this funnellike hole. A colorful patchwork of animals and encrusting algae blankets the cave walls. A long, narrow tide pool runs along the cave floor *(opposite)*.

15 Flat limpets, acorn barnacles, stubby goose barnacles, and a few small mussels cling to the ceiling of the cave above the windowlike hole where water rushes out at high tide *(above)*.

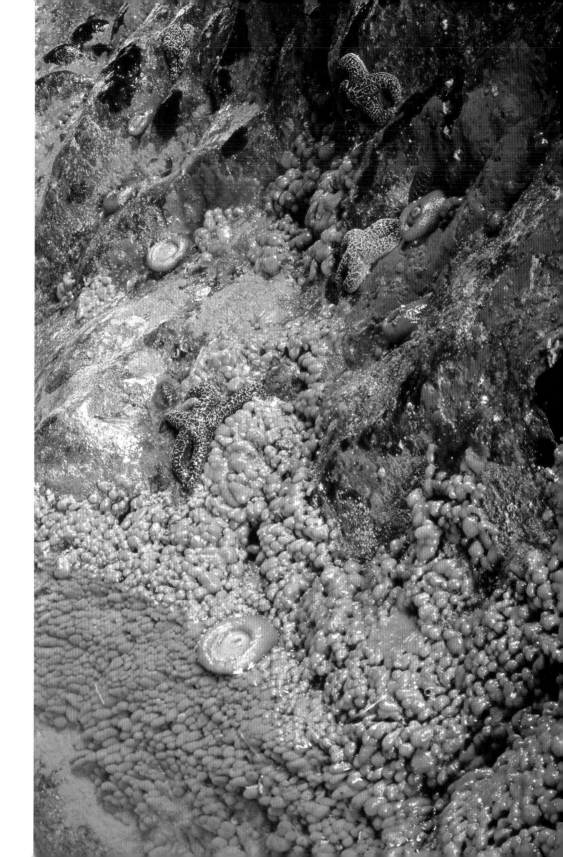

16 Starfish in shades of lavender, rose, or orange cling to the cave walls among crusts of pink coralline algae and colonies of bright red sponges. The fleshy tan lumps and sheets covering so much of the space on this wall are sea squirts, our closest relatives in the cave. (Although these adult sea squirts look like wads of chewing gum stuck to the rocks, as larvae they were little tadpoles, wiggling their tails as they swam through the water.) Sea anemones exposed to air at low tide have retracted their tentacles into their olive-green bodies, which sag like breasts from the wall *(opposite)*.

17 A tiny hermit crab scampers across a red sponge below an awning of white sea anemone tentacles. The crab carries its house, the shell of a dead snail *(above)*.

18 Looking straight down at the floor of the cave, peering into the dark recesses of a pool deep at the base of the sloping, life-encrusted walls, we see the pale faces of a few eerie, faded sea anemones among the loose rocks.

with fleshy creatures like those we saw deep under rock overhangs and in tide pools outside the cave. Purple and red sponges and lumpy mats of soft tan sea squirts grow across the rock. Solitary starfish and sea anemones dot this colorful, soggy quiltwork of animals. Bend over and peek into spaces between the obvious creatures to discover the many smaller animals and plants living there. Hermit crabs and other little crustaceans scuttle around the sea anemones and sponges glued to the rocks (plate 17). Turn your eyes to the bottom of the cave to glimpse the ghostly white sea anemones lurking in the dark recesses of a pool deep in the cave floor (plate 18).

Where Are We Going?

This quick tour, in words and pictures, of one small section of a rocky coastline illustrates that by looking closely at any patch of the natural world around us, we can find within it a rich collection of microhabitats. The animals and plants living in different microhabitats within one site can encounter physical conditions quite different from those experienced by their neighbors just a few centimeters or meters away.[2] For this reason, if we want to learn how a living thing interacts with its environment, we must observe the habitat from that animal or plant's point of view. If a

---

2. In this book we use the International System of units of measurement (called *SI units*). These are the units that scientists use, and they are also the measures used for commerce in many countries around the world.

The SI unit of length is the *meter,* which is roughly a yard. A *centimeter* is one hundredth of a meter; one inch is about two and a quarter centimeters long. A *millimeter* is one thousandth of a meter, about the width of a poppy seed. A *kilometer* is one thousand meters long, which is approximately the length of ten American football fields.

The SI unit for volume is the *liter,* which is just a little smaller than a quart.

(note continues on page 24)

creature lives in a narrow crack, we need to measure what the conditions are like in that crack in order to understand how the tiny beast functions in nature.

This book focuses on just one stretch of shore to emphasize the surprising diversity of microhabitats that can be found within a single site if you know how to look for them. But we also hope to convey, through our photographs and prose, a deeper sense of this magnificent place, so extraordinarily rich in shapes and colors, in textures, movement, and change. It is a feast for the senses as well as the brain. For us, the beauty of this place is enhanced by a growing understanding of how it works, and our studies of how it works are made all the more exciting by the stunning setting in which the processes play out.

How and Why

This is a book about *how* plants and animals of different body designs live where they do. It is not a book about *why* they have those body designs, which is the result of evolution. It is also not a book about *why* they live where they do, which is the result of a variety of ecological processes.

The beautiful and bizarre living things we find on the shore were sculpted into their present forms by the process of evolution. Many books

---

The SI unit for mass is the *kilogram*. A liter of water has a mass of one kilogram, which is a little less than the mass of something that weighs a half pound.

The SI unit for force is the *newton*, which is roughly as much force as the weight of three apples.

The SI unit for time is the *second*, so we express speed as *meters per second*. One meter per second is a little faster than two miles per hour.

about seashore life (see the "Additional Reading" section) are natural history guides that help us identify these weird, wonderful animals and plants. Some of these texts also delve into why the animals and plants look the way they do by tracing their family trees and explaining their evolutionary history. In contrast, our book will examine how the structure of a plant or animal affects its performance of specific functions but will leave the issue of why it happens to have that structure to the books about evolution.

A number of different processes determine where particular types of plants and animals live on the shore. Of course, the places in which a living thing can survive are limited by the physical conditions it is able to tolerate. However, many factors in addition to the physical environment also affect where particular species thrive. For example, interactions with other living things can determine which species are abundant at a site. Members of a particular species may be rare at locations where they are mowed down by predators but can be numerous in other habitats where those predators are absent. At some spots on the shore, members of one species might outcompete other types of plants and animals for space on the rock, while in places where those superior competitors are done in by predators or local physical conditions, the other plants and animals can be abundant. In addition, the abilities of various types of living things to reproduce, disperse to new locations, and colonize different habitats can influence where they live. Furthermore, destructive events such as storms, fires, oil spills, and human construction projects can have a profound effect on which species occupy a habitat. The interaction of all these processes produces the spatial patterns of living things that we see at a particular site. Because so many marine ecology books already discuss why plants and animals live where they do, in this volume we will take a different approach. We will simply meet living things where they occur and ask how they cope with the physical conditions at that place.

Thus, our goal in *Wave-Swept Shore* is to complement the many books now available about the evolution and ecology of seashore life with a different sort of book—a book about the physical environment experienced by plants and animals on the shore and how they cope with it. We will examine the water movement, mechanical forces, heat, light, and humidity encountered by living things in different microhabitats on the coast, and we'll look at some features of the plants and animals that enable them to function in such conditions.

Intuitive Biophysics

This is a book about the biophysics of animals and plants interacting with the environment. *Biophysics* is simply the physics of living things. Many biophysicists study the physical behavior of biological molecules, while others investigate such topics as how cells change shape or how animals move about. *Biophysical ecologists* study physical conditions in the environment, the effect of living things on those conditions, and mechanisms that plants and animals use that enable them to survive the physical rigors of the natural world.

The equations and specialized terminology that scientists use to communicate with each other within their disciplines provide a precise and powerful language for describing how the natural world works. Nonetheless, we have chosen to tell the story of how plants and animals cope with the rigors of a wave-swept shore without using species names or anatomical jargon to describe them, and without using mathematics to describe the physics. We rely instead on photographic images of a rocky shore and your experience of the physical world to explain how living things interact with their surroundings. The principles discussed in this book can be grasped by anyone who is willing to observe nature carefully and to pay close attention to the physical phenomena we experience every day.

As you read this book, be mindful not only of the place of a living being in the environment, but also of its size.

2

WATER AND LIFE ON THE SHORE

Why water? This book is about water and what it does to and for the living things on the shore. Animals and plants glued to coastal rocks risk being torn away by crashing waves, and yet they depend on the water moving around them to bring them oxygen and nutrients, to carry away their wastes, and to transport their offspring to new sites. When stranded in the air as water recedes at low tide, aquatic plants and animals on the shore may perish if they become too dry. Out of the protective bath of the sea, inhabitants of the intertidal can be cooked by the heat of a summer day or frozen in the chill of a winter night. The tumultuous motion of coastal waters and the rise and fall of the tides have a profound impact on the lives of plants and animals on the shore.

The Force of Moving Water

As anyone who has tried to stand his ground in the surf knows all too well, moving water imposes forces on a body that might sweep it away. As waves batter coastal rocks, the algae and animals glued to their surfaces can be deformed, torn apart, or ripped from the substratum.[1] Unattached animals like crabs risk being buffeted about and washed away. In chapter 3 we examine how water moves near the shore and how living things can affect the speed of the flow hitting them. Then in chapter 4 we see how moving water imposes forces on the bodies in its path. Two creatures exposed to the same water speed can feel forces of different magnitudes if their sizes or shapes are not the same. Furthermore, diverse living things bearing forces of the same magnitude can experience different fates: some stretch or bend, some break, and some come unstuck from the rocks. We borrow approaches employed by engineers to analyze man-made structures and use them to figure out how the architecture of different animals and plants on the shore affects what happens to them in waves.

Transport by Moving Water

Not only does moving water impose forces on the inhabitants of the shore, but it also carries with it dissolved substances and suspended objects. Currents transport these things from one place to another. The ocean is turbulent, with eddies[2] large and small swirling around in the currents as they move along (plate 20). These turbulent eddies stir the water, mixing and spreading waterborne materials in the same way that the cream gets dispersed in your coffee cup when you stir it. If you pour some milk into a stream or into the ocean, you'll see the white blob of milk grow larger and more diffuse as turbulent

---

1. The *substratum* is the solid surface on which an organism lives.

2. An *eddy* is a vortex of fluid swirling around in a circle.

20 Swirling vortices in the water rushing through channels between the rocks mix things up. These eddies stir depleted water and wastes away from the animals and plants clinging to the shore *(opposite)*.

eddies mix up the milk with the surrounding water, and you'll also see that this expanding, diluted blob of milk is carried away by the current. The same thing happens to materials released by animals and plants on the shore.

Marine plants and animals take up oxygen, dissolved nutrients, or food from the water around them and release wastes, such as ammonia, carbon dioxide, and feces. If the water around an animal or alga is still, it can become depleted of the substances that the individual uses and fouled with its wastes. In contrast, if the water is flowing, turbulence stirs away the halo of depleted, fouled water around the individual, and currents carry off stale water from the neighborhood, replacing it with new water from elsewhere.

Just as some terrestrial plants rooted to one spot rely on wind to disperse their pollen and seeds, many marine plants and animals stuck to the bottom depend on the water flowing around them to transport and mix their sperm and eggs and to disperse their offspring to new sites. If adults of certain species are ripped from the shore, they can raft in the moving water and colonize another spot on the coast.

Flowing water also transports inanimate objects like sand grains and rocks. The local speed of the water across particular areas of the shore determines where sediment is swept away and where it is deposited, sometimes burying animals and algae. Logs and rocks hurled by waves, and boulders rolled along the bottom by the surf, smash and scrape the shore, cutting swaths of destruction through the mats of life on the rocks.

Chapter 5 explores how water transports these substances and objects so important to the living things on the shore. We'll examine how the turbulence of near-shore flow stirs things up, and how waves and currents move materials around coastal areas. We'll also see how various types of algae and animals take advantage of the transport service provided by the ocean.

21 Moving water transports sand and rolls rocks.

22 Sea anemones in a small pool and clumps of algae in shady crevices are
more protected from drying in the sun than they would be on the bare, flat
rock around these refuges.

The Presence or Absence of Water

Marine creatures can dry up, cook, or freeze when not protected by surrounding water. Furthermore, many saltwater plants and animals cannot tolerate fresh water and will perish if bathed in rain. When the tide goes out, some shoreline inhabitants are exposed to sun, wind, and rain, while others live in microhabitats where they are protected from these dangerous terrestrial conditions (plate 22). For example, intertidal plants and animals stranded above the waterline as the tide recedes can be sheltered in pools, under rock overhangs, in crevices and caves, or within dense canopies of seaweeds or mussels, while those on the exposed faces of wave-battered rocks can be showered with spray. In general, creatures high on the shore spend more time exposed to the perils of the air than do those lower down on the rocks.

Chapter 6 considers the physical conditions in different types of microhabitats on the shore when the water recedes at low tide. We examine how factors like wind speed affect the dangers of drying, stewing, or freezing, and we look at the diverse ways in which intertidal animals and plants cope with the rigors of exposure to air.

Water Is Essential for Life, and Water Is Dangerous

Water motion plays so many critical roles in the lives of the inhabitants of the shore that we will focus our book on this important aspect of their physical environment. How does water move at our wave-swept shore, how do the plants and animals living there interact with it, and how do they survive when stranded without it at low tide?

3

WAVES

Flow Felt by Living Things on the Shore

As we stand at the coast, hearing the thunder of wave on rock, watching the geysers of spray, it is easy to imagine that the creatures clinging to the shore are being battered by enormous forces. However, by carefully watching how water flows in waves moving across the complex terrain of a rocky coast, we learn where refuges of calm develop. A stretch of shore that appears to be exposed to a maelstrom of rushing water can actually harbor many animals and plants that are protected from the brunt of the waves.

Water Movement in Waves

When we stare out to sea and watch waves moving across deep water, before they arch over and break on the rocks, we can be mesmerized by the vision of wave upon wave rolling relentlessly across the ocean towards the shore. Watching carefully, however, we can see that pieces of debris carried in the

23 Wave shapes move across deep water, but as they near the shore, where the sea is shallower, the ocean bottom slows their shoreward motion. As the waves slow down, they grow taller. They lurch toward the land and finally topple over, breaking in a spectacle of spray and foam *(opposite)*.

water as these nonbreaking waves pass do not move toward the shore at the same speed as the waves. The debris reveals how the water itself is moving when the wave shape passes by. At the crest of a wave, water flows in the same direction as the wave is travelling, but in the trough, water moves in the opposite direction. Water moves up as a wave crest arrives, and down as the crest travels beyond it. By watching debris we learn that even though wave shapes sweep toward the shore, the water does not. Instead, the water flows around in vertical circles, only moving slightly shoreward with each passing wave, one step forward, half a step back. Eventually the debris and the water that carries it make their way to the shore.

If you go snorkeling just under the water's surface in nonbreaking waves, you can watch how particles suspended in the water move around when the waves roll by. The water does indeed move around in vertical circles—up, shoreward, down, seaward, over and over. (You move the same way if you let yourself float passively as the waves move by.) The deeper you swim down below the surface, the smaller the orbits of water motion become, and eventually they disappear. In deep water, living things on the bottom do not experience the motion of waves passing overhead (see diagram 1).

As waves move shoreward and travel across shallower water, their behavior changes. A rule of thumb for defining shallow water is that the depth is less than half the crest-to-crest distance between successive waves. In shallow water, waves "feel" the bottom, and the circular orbits of the water in waves are squashed into elliptical paths near the sea floor. In fact, a creature hugging the bottom at a shallow site encounters back-and-forth water flow, called surge, as waves pass over it. The water moves shoreward under wave crests and seaward under the troughs. The bigger the waves, the faster the flow. In addition, the bigger the waves overhead,

A

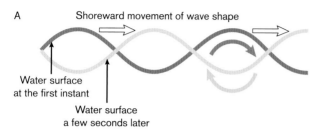

Shoreward movement of wave shape

Water surface
at the first instant

Water surface
a few seconds later

B

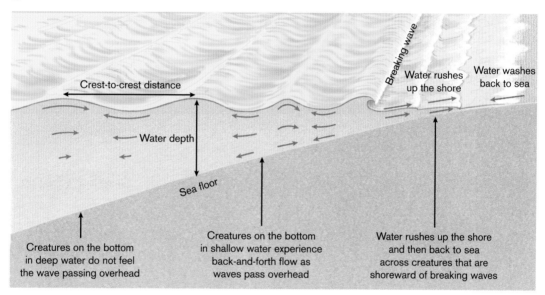

Crest-to-crest distance

Water depth

Sea floor

Breaking wave

Water rushes
up the shore

Water washes
back to sea

Creatures on the bottom
in deep water do not feel
the wave passing overhead

Creatures on the bottom
in shallow water experience
back-and-forth flow as
waves pass overhead

Water rushes up the shore
and then back to sea
across creatures that are
shoreward of breaking waves

DIAGRAM 1 Water motion in waves. (A) We see some waves at one instant shown
in dark blue, and then those same waves a few seconds later shown in light blue. The
white arrows show the direction that the wave shapes have moved between the two instants. The dark blue arrow shows the direction the water is moving under a wave crest
at the first instant, and the light blue arrow shows the direction that same bit of water
is moving at the second instant. If you watch one patch of water over time, you see that
it flows around in a vertical circle, shoreward under the wave crest, then down as crest
moves away, then seaward under the wave trough, and then up as the next wave crest
arrives. (B) During one snapshot in time, we see the directions that the water is moving
at different positions near the shore.

24 After each wave breaks, water rushes up the shore, slows down, and then flows back into the sea. This back-and-forth flow of water happens over and over, wave after wave.

the longer the water moves in one direction before it turns around and flows back the other way across the plants and animals on the bottom.

When waves move across shallow areas, the waves are slowed by their interaction with the sea bottom. As they slow down, they grow taller and eventually become unstable. The water at the wave crest lurches shoreward and topples downward, and the wave breaks (plate 19). The water from a wave that has broken rushes across the shore, slows down, and then washes seaward again. Intertidal animals and plants that sit shoreward of breaking waves encounter this back-and-forth swash of water. Water velocities and accelerations in breaking waves and in the back-and-forth flow of broken waves can be huge.

When we stand on a cliff overlooking the shore, hearing the roar of wave after wave crashing across the rocks, we can be awed by imagining the physical brutality of life on those rocks. Are we right? The most direct way to find out would be to leave our dry perch and climb down onto the shore to measure just how rapidly the water sweeps across different plants and animals. Some scientists are actually crazy enough to do this. We slither and climb across algae and barnacles at low tide, lugging a gasoline-powered masonry drill, so that we can sink bolts into the rocks. We use the bolts to attach electronic flow-measuring probes to the shore right next to specific animals and plants. Retracing our steps back to dry land, we lay waterproofed wires from the probes across the shore, carefully anchoring them as we go. We string the wires up to a safe perch and plug them into a laptop computer; then we sit and wait for the rising tide to submerge the probes. On lucky days, when the rain does not drown our computer, salt does not short out our battery connections, and waves don't wash our cables loose, we record the water velocities actually encountered by specific plants and animals on the shore. Our measurements show patterns of water flow that make sense to anyone who has watched waves carefully: water movement experienced by living things on the shore is faster when the water is shallow, when the waves are big, and when the waves break nearby. However, our measurements of water velocities where the animals and plants live also reveal some surprises. The local water flow experienced by individual creatures or algae can be quite different from the mainstream flow overhead. In fact, the water movement encountered by one individual can be much faster or slower than the flow experienced by its next-door neighbors, only a few millimeters or centimeters away on the very same rock!

Air Flow and Water Flow

We terrestrial animals can begin to understand how water moves through the complex texture of a rocky shore if we think about how wind moves around structures on dry land. Both liquids (like water) and gases (like air) are fluids. *Fluid dynamics* is the study of how liquids and gases move: *hydrodynamics* is the study of how water moves, and *aerodynamics* is the study of how air moves. Biologists can use fluid dynamics to investigate how living things interact with the water or air swirling around them in nature.

We can understand what fluids are by contrasting them with solids. A solid material, like rubber or steel, resists being deformed by the force you impose on it. Imagine trying to squash a rubber eraser. The farther you push it down, the harder you have to push. Solids care about *how far* they have been deformed. If you stop trying to mash the eraser, it bounces back to its original shape. Contrast this behavior with the response of a fluid, like water in the bathtub, as you deform it by stirring it with your

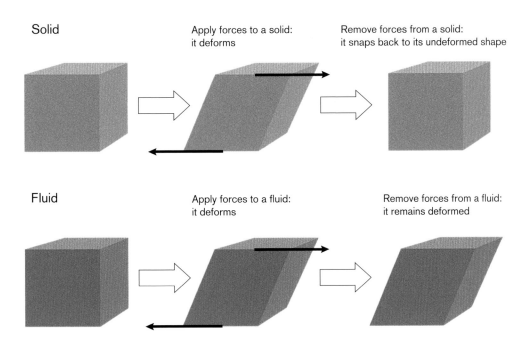

Solid

Apply forces to a solid:
it deforms

Remove forces from a solid:
it snaps back to its undeformed shape

Fluid

Apply forces to a fluid:
it deforms

Remove forces from a fluid:
it remains deformed

DIAGRAM 2 When forces (shown by the black arrows) are applied to a solid and a fluid, both deform, but when the forces are removed, the solid snaps back to its original shape (unless it was damaged), whereas the fluid remains deformed.

hand. It is much easier to push your hand through the water if you move it slowly than if you force it quickly. Fluids care about *how rapidly* they are deformed, but not about how far they have already been deformed. When you stop moving your hand, the water does not act like the eraser: it does not snap back to where it was before you started deforming it. Air is also a fluid, so the faster you try to move it, the harder you have to push, and air does not bounce back to where it used to be when you stop pushing (diagram 2).

Since both air and water are fluids, our intuition about how air moves around us and exerts forces on us helps us understand how moving water pushes on marine animals and plants. However, we have to keep in mind one big difference between water and air. How hard a moving fluid pushes on a body standing in its path depends on the density of the fluid. Density is the weight of a measured volume of the fluid: a liter (about a quart) of water weighs a thousand times more than a liter of air. Water pushes on you with a force that is a thousand times greater than a breeze blowing on you at the same speed. Anyone who has tried to stand her ground in a rushing river knows all too well the force of moving water!

Water Flow in the Neighborhood

The terrain of a rocky shore is complex. Channels of water thread their way between massive stone faces and a hodgepodge of boulders. The rocks are carved with cracks, crevices, and caves. Some surfaces wear thick coats of algae or mussels, while others are bald, save for a few wartlike barnacles. As water flows through this rich patchwork of obstacles and textures, it is slowed down, speeded up, and redirected. Water motion encountered by an individual living thing depends on its location in this complex terrain and on its size relative to the bumps, valleys, and other animals and plants nearby. The animals and algae clinging to the shore come in many sizes, from kelp as long as evening gowns to snails the length of a coffee bean and baby barnacles about as tall as a poppy seed. The water flow actually hitting an individual plant or animal on the shore can be very different from what we would expect from looking at the large-scale water movement over the site.

We can understand how specific features of the topography of a shore affect the water flow experienced by plants and animals living there if we

25 At low tide, the channels winding their way between rocks on the shore vary in width from gaping gaps to skinny passages. As the tide rises and water courses through this complex network of spaces, its paths are altered and its speeds are changed.

26 A breaking wave crashes into rocks on the shore.

think about how we find shelter from the wind on land. One way to get out of the wind is to hide on the lee (downwind) side of a structure, like a wall or a shed. Similarly, animals and plants sheltered behind a rock encounter slower water flow than do those on the upstream side of that same rock. This difference in local flow is obvious if you watch the movement of bubbles carried by the water as it rushes across the front of a big boulder and then swirls in slow eddies behind it. Unlike wind, however, water in waves moves back and forth. When waves pass over a boulder, the upstream and downstream directions are reversed every few seconds. Even so, living on the seaward side can be more dangerous. At times in the tidal cycle when waves break right on a rock, the seaward face of the rock sitting in air can suddenly be slammed by a wall of water as a wave crashes into it (plate 26).

Another way to escape the wind on land is to stand downwind from your buddy or to huddle in the middle of a group of people. Those on the perimeter of the group feel faster wind than you do as you hide, sheltered by them, in the center of the cluster. The poor souls on the upwind side of the group bear the brunt of the gale. The same thing is true for aggregations of living things on the shore, such as beds of mussels, mats of sea anemones, clusters of goose barnacles, and clumps of algae: the individuals in the middle of such a group are sheltered from rapid water flow by their neighbors (plate 5).

If you are alone, the tallest thing around as the wind whips across an open field, you can escape the blast by crawling down into a dry streambed cut into the ground. In the same way, a crab hiding in a crevice as waves rush over a rock is protected from the torrent.

You can retreat from a windswept field to find refuge in a copse of trees or a nearby woodland. The air moves slowly through the canopy of plants around you as the wind skims over the tops of the trees. The denser the vegetation, and the more acreage covered by the forest, the slower the breeze within it. A small animal, like a cat, can find such shelter within a clump of bushes; a mouse and an ant are even protected from the wind under a canopy of grass. In the same way, water flows around and over canopies of marine plants or animals. Flow is slow within kelp forests, surfgrass meadows, mussel beds, and algal turfs. Such canopies provide wave-sheltered habitats for a rich diversity of living things.

Continuity in Caves

You can avoid the wind altogether by staying indoors. Are the animals hidden away in a sea cave also "indoors," protected from the wrath of the waves? Because each cave has a unique configuration, we must determine the answer to this question anew for every cavern we explore. The sea cave on our particular stretch of shoreline (shown in plates 13 and 14) is like a cabin with its front door flung wide open to the oncoming wind, because its entrance faces seaward into the oncoming waves. The mouth of the cave opens into a wide chamber, like the living room inside a cabin. At the far end is a narrow, window-like opening. When a big wave breaks outside our sea cave, the water rushes in through the mouth, flows more slowly through the wide chamber, and then blasts out through the nozzle-like back window (plate 27, diagram 3). When nonbreaking waves roll through our cave, the water flows back and forth as it does outside the cave, but the water moves more slowly across the rock surfaces in the wide midsection of the cavern than it does over the rim of the narrower mouth or the sill of the small window. (In other sea caves that are blind pockets, with no shoreward exit, the water that rushes into the cave must slosh back out again through the mouth.)

27 If we stand atop the cliffs and look down at the shoreward side of the
rock island where our cave is located, we have a good view of the rear window of the cave. When a wave crashes into the seaward side of that island,
water rushes into the mouth of the cave, flows through the cavern, and then
shoots like a geyser out the narrow rear window. The view of this window
from inside the cave shown in plate 14 reveals just how skinny this exit
nozzle is compared with the spacious interior of the cave.

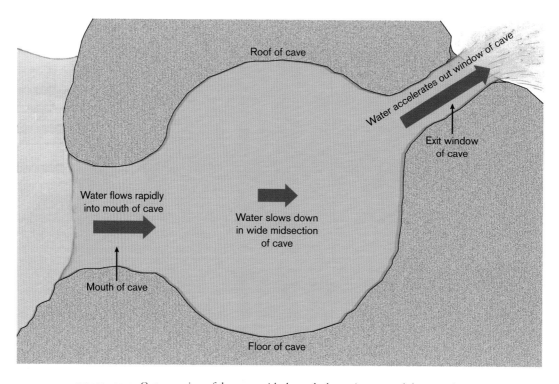

Roof of cave

Water accelerates out window of cave

Exit window
of cave

Water flows rapidly
into mouth of cave

Water slows down
in wide midsection
of cave

Mouth of cave

Floor of cave

DIAGRAM 3 Cutaway view of the cave, with the rock shown in gray and the water in blue. The longer the blue arrows, the faster the water flow. The seaward-facing mouth of the cave is to the left. Water flows into the cave mouth, slows down in the wide midsection of the cave, and then accelerates and flows very rapidly out of the narrow exit window at the back of the cave.

To understand why the velocity of the water changes as it moves through different sections of an enclosed space like a cave, think about the garden hose you use to wash your car. When you turn on the tap, water flows through the hose and out of its end at a given rate. You can measure that rate by timing how long it takes to fill a bucket. Let's say that you have a flow rate of two bucketfuls per minute. If you want to blast

28 Fragile red sponges and flabby beige sea squirts cover the walls of the
midsection of the sea cave. A flat chiton, with a row of shell plates along its
back, creeps across the rock. (Chitons are mollusks, in the same phylum as
mussels and snails.)

29 Robust ranks of goose barnacles are plastered on the ceiling near the constricted exit from the cave.

the bugs off your windshield, you simply hold your finger over part of the opening at the end of the hose. The stream of water then jets out of the small opening that you have left at the end of the hose and gushes onto the windshield at a high speed. Two bucketfuls of water still have only a minute to funnel through the constricted opening at the end of the hose, so the water has to speed up as it flows through the narrower hole. A skinny nozzle screwed onto the end of the hose creates the same effect. Your garden hose illustrates a general rule called the *principle of continuity:* when fluid flows through a pipe at a given volume per time, it travels at higher velocities where the pipe is slender and at slower speeds where the pipe is wide. That is why water moves rapidly through the mouth of our cave, slows down in the wider interior of the cavern, and then speeds up again as it is funneled out through the narrow exit window. We should not be surprised to find delicate, crumbly sponges and soft sea squirts carpeting the walls of the wide interior of our cave where the flow is slow (plate 28) but dense mats of tough goose barnacles and armored mussels around the nozzle-like exit, where the water rushes by rapidly (plate 29).

If you are a city dweller, you also experience the principle of continuity on windy days, as you are buffeted by the rapid rush of air funneled through the narrow canyons between tall buildings. In a similar way, water speeds up as it sweeps through the gaps between boulders and rushes through narrow channels that dissect the rock benches.

Refuge in the Boundary Layer

When you are out on an open plain or a flat, sandy beach with no place to hide, you can still get out of the wind by hunkering down close to the ground. If you sit, the wind on your face is slower than when you are standing, and if you lie down it is slower still. When you flatten yourself on the ground and experience the slower air movement there, you are sheltering within a region that fluid dynamicists call the *boundary layer.* Whenever a fluid (like air or water) flows across a solid surface (such as a beach or a boulder), the layer of fluid actually touching the solid does not slip along its surface but rather just sits there, not moving. This fluid stuck to the surface hinders the movement of the layer of fluid right above it, and that slowed layer in turn resists the flow of the fluid above it, and so on. The higher above the surface you look, the faster the fluid moves along, until you reach the layers of fluid that are not retarded by the bottom; here you are outside the boundary layer and you are subjected to the *free-stream flow* (see diagram 4).

Remembering your experience of the boundary layer along a windy beach, look again at the living things on the shore. Flat animals and plants plastered onto rock surfaces are also hiding in the slowly moving water of the boundary layer. Examples of this encrusting body form can be found in many different groups of animals, from our sea-squirt relatives to primitive sponges that look like colorful splotches of paint in wild abstract murals on the stone walls of the shore (plate 30). Crustose algae share the refuge of the boundary layer, some resembling chalk drawings in shades of pink and purple and others forming rubbery black coatings on the rock (plate 31).

Like encrusting creatures, tiny animals and plants hugging the bottom encounter slower flow than their taller neighbors. How much faster does the water movement get if a plant or animal grows a bit taller? The answer depends on a number of factors that we can observe and measure. The faster the free-stream flow across a surface, the thinner the boundary layer. Therefore, standing up even just a little taller has more striking consequences for an inhabitant of a rock exposed to rapid water move-

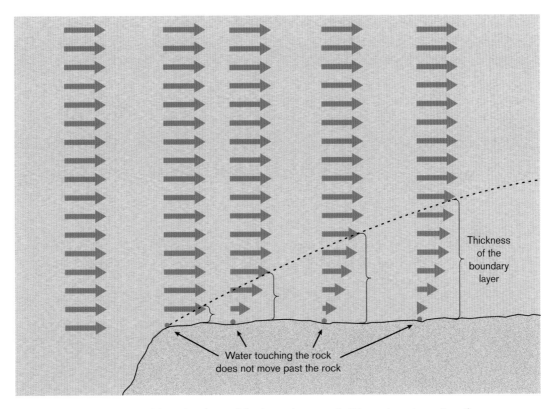

DIAGRAM 4 A boundary layer of slowly moving water builds up along the surface of a rock as water flows across it. The longer the blue arrows, the faster the flow. The dots indicate a velocity of zero meters per second where water contacts the rock and does not move past it.

ment than for a resident of a more protected spot. Furthermore, if the flow is very turbulent, eddies can swirl down into the boundary layer and sweep away some of the slowly moving water close to the bottom. Thus, the more turbulent the water, the shorter a living thing must be to find refuge from rapid flow in the boundary layer.

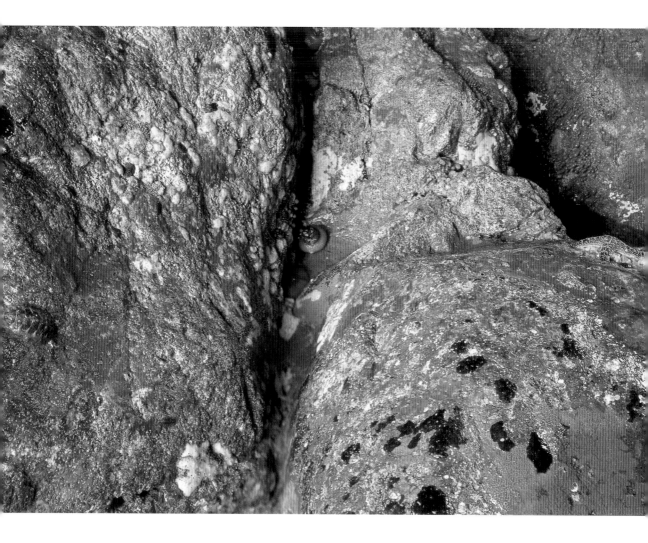

30 The walls of the cave are encrusted with a colorful array of flat animals and algae. Sponges in dazzling shades of red, orange, and yellow contrast with the creamy beiges and whites of the fleshy sheets of sea squirts. A background of chalky lavender coralline algae, accented by splats of shiny black crustose algae, completes the picture. A few big blobs, sea anemones, lurk in the crevice cutting through this living mural *(opposite)*.

31 Rubbery black algal crusts look like tar stuck to the rocks. These flat algae encounter the slow water movement that occurs right next to the surface of the rock, as do the tiny black snails and short, turfy seaweeds alongside them *(above)*.

57

Plants and animals on the upstream edge of a big rock have to be shorter to hide in the boundary layer than do individuals living farther downstream along the same rock. The reason for this difference is that it takes a while for a boundary layer to build up. When flowing water first runs into the upstream end of a rock, the layer of water right next to the rock is slowed down, and there is a big difference between its retarded pace and the speed of the water rushing along swiftly right above it. As the water moves beyond the leading edge of the rock, the layer of slowed water right next to the rock surface then retards the movement of the layer of water just above it. A bit farther along, that next layer of water begins to hinder the progress of the layer above it, and so on. Thus, as the water flows farther and farther across the rock, the boundary layer of slowed water builds up thicker and thicker.

Remember that the water flows back and forth along the bottom under waves. It takes some time for the boundary layer to build up as the water flows in one direction. Then, when the water flows back the other way, the process starts all over again, and the boundary layer has to build up once more. Therefore the boundary layer is thinner in wavy flow than it is in a current moving at the same speed in just one direction. That means that plants and animals have to be shorter to hide in the slowly moving water within the boundary layer at a wave-swept site than they do to hide in the boundary layer at a site exposed to a steady one-way current.

Large and small sea anemones sitting side by side offer a vivid example of how the heights of creatures can determine the speed of the water movement they experience. The big green sea anemones dotting our stretch of rocky coastline stand taller than their diminutive, pink-tentacled neighbors, sea anemones of another species (plate 32). We measured

32 Big green sea anemones look like giant olives when they retract at low
tide. They are surrounded here by smaller, pale anemones of another species
that can reproduce by splitting in half longitudinally to form mats of tightly
packed individuals covering the rocks. When the tide comes in, the large
anemones inflate their vibrant green sunlike faces (plate 50), while the small
sea anemones expand to form a low carpet of delicate pink tentacles. The
taller green animals encounter water movement ten times faster than the
flow that their short pink-tentacled neighbors experience while hiding in
the midst of a dense crowd of their sisters.

the velocity of the water rushing through a channel cutting between rocks on a wave-swept shore. The bottom of the channel was carpeted by both species of sea anemones. We discovered that the water was moving about a hundred times faster in the middle of the channel than it was down close to the bottom where the squat pink-tentacled anemones lurked, but only ten times faster than the flow hitting the emerald tentacles of the big anemones sitting right next to them. Next-door neighbors on the shore don't necessarily experience the same flow, and a place that looks hydrodynamically harsh to us may be calm for the local inhabitants.

Some animals can change their height in response to local flow conditions. When big waves are crashing, the large green sea anemones flatten themselves into pancakes only two to three centimeters tall (shorter than the diameter of a golf ball), but when water motion is more gentle, these animals inflate themselves to heights of ten centimeters or more. Surprisingly, the speed of the water past their tentacles under such very different flow conditions turns out to be about the same. You can experiment with this phenomenon yourself: the air flow across your face as you lie flat on a beach on a windy afternoon can be every bit as slow as the breeze that brushed your brow as you stood tall on that beach a few hours earlier, facing a gentle morning breeze.

Things Are Not What They Seem

A wave-battered shoreline is a harsh, dangerous place for us big-bodied humans. However, if we look closely at this rugged environment, we discover that it harbors a rich array of flow microhabitats. While animals and plants on the seaward face of a rock are hit by crashing waves, others, hunkering down in crevices or snuggled among the seaweeds and mussels, may encounter practically no flow at all. An encrusting alga spread out flat on a rock surface lives in a very different hydrodynamic world from that of the tall seaweed standing next to it. When you look at water flow on the shore, or feel wind moving across the land, focus your attention on exactly where plants and animals of different sizes are living in that seascape or landscape. Consider the flow from their perspective.

4

FLOW, FORCE, AND FRACTURE

Body Architecture

When we venture onto the rocks at low tide, we find squishy sea anemones and floppy seaweeds clinging to the rocks alongside the well-armored barnacles and mussels (plate 33). The sleek, rigid profile of a limpet is very different from the lettucelike form of an alga, yet both of these body architectures function in this wave-swept world. How do such diverse living things withstand the hydrodynamic forces that threaten to wash them away? Biologists can unravel the mysteries of how plants and animals withstand crashing waves by borrowing techniques developed by engineers to analyze the mechanical performance of man-made structures.

How does the shape of a plant or animal clinging to the shore affect the hydrodynamic forces it feels when awash in waves? We can approach

33 Low tide reveals a shoreline festooned with diverse body types. A mat of soft, squishy sea anemones carpets a rock in the foreground. The surrounding boulders are covered with a salad of seaweeds: bushy brown turfs, leathery straplike fronds, and delicate, bright green sea lettuces (*opposite*). These flexible forms contrast with the stiff, hard shells of mussels (plate 6) and limpets (plate 12) clinging to the shore and the rigid, volcano-shaped houses of barnacles plastered to the rocks (plate 11).

this question in the same way that engineers investigate how the profile of a piling or a submarine determines how big the forces are that it bears when water whizzes by. This approach reveals why the force on one creature can be much bigger than the force on its neighbor hit by the same water flow.

Will the force of moving water rip an animal or alga off the shore? To answer this question, we study the architecture of a plant or animal much as an engineer analyzes a beam or a building. We've learned that the shape and size of a plant or animal determine how hard its tissues are pulled or squashed when it is subjected to hydrodynamic forces. Are those tissues stiff or stretchy? Are they brittle or tough? How easily an individual's tissues deform and how readily they break determine the fate of that animal or plant in response to the mechanical forces imposed on it by moving water.

Hydrodynamic Forces on Bodies

Water rushing past an object exerts a hydrodynamic force called *drag* that pushes the body downstream. The faster the flow, the bigger the drag. In fact, drag is proportional to the square of velocity, so doubling the water speed past a body leads to a fourfold increase in drag, tripling velocity leads to a ninefold rise in drag, and quadrupling produces a force sixteen times bigger. This means that if an animal or plant encounters a slightly faster flow, it experiences a huge increase in drag. Conversely, hiding in a microhabitat protected from rapid flow profoundly reduces the drag tending to sweep a plant or animal away.

Drag depends on how much a body disrupts the flow around it. When a big object is exposed to a water current or wind, a turbulent wake forms

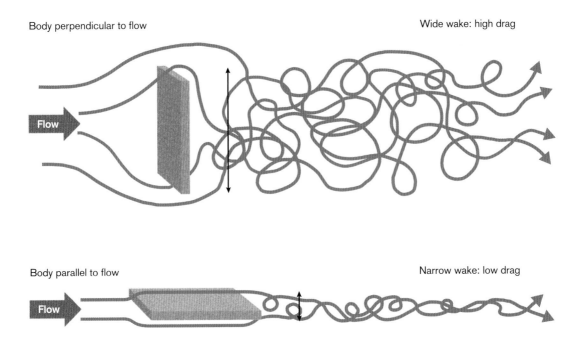

Body perpendicular to flow

Wide wake: high drag

Flow

Body parallel to flow

Narrow wake: low drag

Flow

DIAGRAM 5 The larger the turbulent wake that forms behind an object, the bigger the drag force on that object tending to push it downstream. Above, a body perpendicular to the flow has a wide wake and experiences high drag; below, the same body oriented parallel to the flow has a narrower wake with lower drag.

downstream of it. Such a wake is obvious if you watch the water swirl around and eddy behind a piling supporting a bridge or dock in a river. The wider the wake, the bigger the drag on the object. You can feel this for yourself by sticking your hand into a flowing stream (see diagram 5). The drag pushing your hand downstream is much greater if you hold it broadside to the flow than if you orient it parallel to the direction in which the water is rushing by. As you struggle to hold your hand broadside to the current, you can see a wide, messy wake form behind it; when you turn your hand parallel to the flow direction, the wake is much narrower.

Scientists have measured the drag on a variety of shoreline animals and plants by attaching them to force-measuring devices and then exposing them to water currents or waves, either out on the shore or in a laboratory *flume* (a flume is a trough or channel through which we make water flow at velocities that we control). These experiments have shown us how the size, shape, and orientation of a plant or animal affect the drag it experiences. A simple way to envision how each of these features affects drag is to think about the width of the wake that forms behind a body in flowing water. For example, when exposed to water flowing at a defined speed, big bodies produce wider wakes and experience higher drag than do smaller individuals of the same shape and orientation. Getting a little bigger makes the drag a whole lot worse. For example, in a given water velocity, growing twice as wide while keeping the same body shape increases the drag fourfold, tripling in width raises the drag by a factor of nine, and quadrupling in size ups the drag sixteenfold!

We are familiar with the slender, streamlined shapes of the bodies of fish, birds, and airplanes, with their rounded noses and long, tapering tail ends. A streamlined profile produces a very narrow wake, and thus low drag, when the fluid motion past the body is from nose to tail. If you orient this kind of streamlined body backward in the flow, with its nose pointing downstream, the wake is bigger and the drag higher. Because

34 Many animals and plants on the shore present low profiles to the flow. Therefore, the wakes that form behind the flattened starfish, the squat white barnacles, and the encrusting red sponges and black algae shown here are small. The golden hydroids (see also plate 67) poking up through the sponge are flexible and easily bent down out of the way when water flows past them (*opposite*).

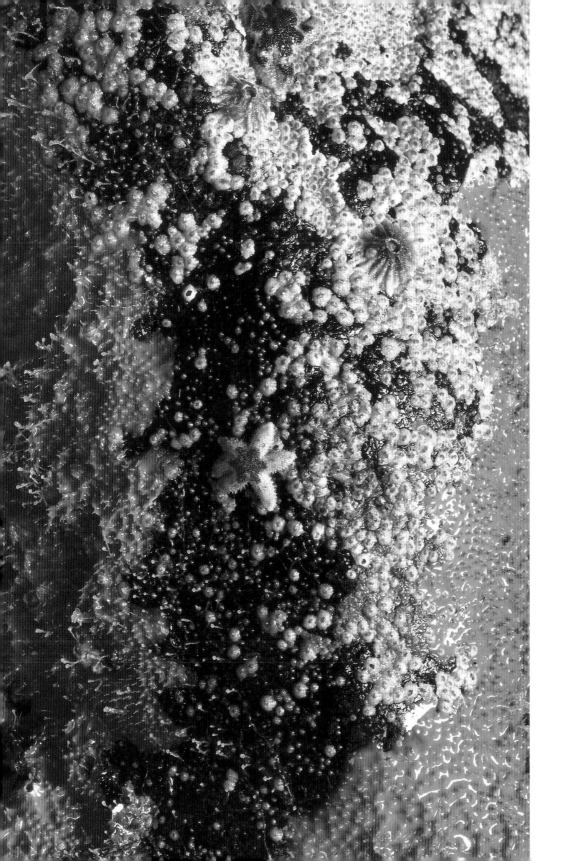

animals and plants living on wave-exposed rocks experience messy back-and-forth water motion, traditional streamlined shapes are not well suited to this environment. Body shapes that produce small wakes no matter which way the water is going do a better job of reducing drag in wave-swept environments than do traditional streamlined shapes. Many of the animals hugging a wave-swept shore have relatively flat bodies with tapering edges all around. Examples include sleek limpets (plate 15), flat chitons (plate 28), squat barnacles, and pancake-like starfish (plate 34). Likewise, encrusting sponges and algae have low, flat, drag-reducing profiles (plate 34). Not only is the drag on a flat body oriented parallel to the flow lower than the drag on a spherical body experiencing the same water velocity, but a flat animal or plant plastered to the substratum in nature also encounters lower water velocities within the boundary layer than a taller neighbor experiences.

As water sloshes back and forth in waves, the flow velocity, and hence the drag force, on a wave-swept plant or animal is constantly changing from one instant to the next, pushing it first shoreward and then seaward, over and over again. When water changes its velocity it is accelerating or decelerating. Bodies in accelerating or decelerating flow experience another force in addition to drag, the *acceleration reaction force.* At instants when the flow is speeding up, the acceleration reaction force acts in the same direction as the drag, so we add these two forces together to figure out how hard the moving water is pushing on a plant or animal. If the flow is slowing down, however, the acceleration reaction force acts in the opposite direction from the drag, and the net force on the body is reduced. As with drag, the less a body interferes with the flow, the lower the acceleration reaction force. Therefore, flat plants and animals oriented parallel to the direction of water movement experience lower acceleration reaction forces than do big, spherical ones or flat ones oriented broadside to the flow.

A little increment in growth can lead to a huge increase in the acceleration reaction force that a plant or animal experiences in waves. If a body doubles in width while keeping the same shape, the acceleration reaction force also goes up by a factor of eight, and if it triples its width, the acceleration reaction force gets twenty-seven times bigger! Remember that drag also increases a lot as a living thing grows. To make matters worse, a big beast or alga is also more likely than a small one to encounter high water velocities and accelerations in nature. Therefore, body shapes that produce small wakes should be more important for large animals and plants than for small ones.

When fluid flows around a plant or animal with an asymmetric shape, the flow may speed up more to move around one side of its body than around the other. For example, when air moves around an airplane wing, the air flowing over the curved top of the wing moves faster than the air traveling along the flat bottom of the wing. This difference in fluid speed produces *lift,* a force tending to suck the body toward the side experiencing the more rapid flow. Lift acts at right angles to the flow direction, and thus can suck an individual either sideways or up off the substratum. Like drag, lift depends on the square of velocity, so a small increase in water speed causes an enormous rise in the lift on a body. Lift forces on animals with domed upper surfaces, like limpets and barnacles (plates 12 and 34), can be higher than drag. Surprisingly, these animals are in more danger of being sucked off the shore vertically than of being pushed off horizontally by swift currents!

Flexible Flapping Fronds

While barnacles and limpets sit stiffly in one spot, rigidly resisting the waves, floppy algae and seagrasses flap and flail as the moving water pushes them around (plate 35). When flowing water pushes on flexible structures like seaweeds, they are bent over parallel to the flow direction and their branches collapse together into slim, streamlined bundles. Not only do these compliant plants become more streamlined when water moves past them, but they are also pushed over close to the bottom, where flow velocities are low and thus hydrodynamic forces are reduced. However, if algal blades flutter in the flow like flags, the wake that forms behind them can be wider (and thus the drag forces can be bigger) than if they simply stream out parallel to the flow without flapping.

In the back-and-forth flow of the waves, flexible plants and animals move to and fro *with* the water. Does this ability to go with the flow reduce the hydrodynamic forces that the holdfast of a floppy seaweed must resist? (The *holdfast* of an alga is the structure that attaches it to the rock; see plate 36). We can find out by affixing a kelp to a force-measuring device bolted to the shore. While the alga is moving along *with* the water and its stipe (or stem) is slack, no force is transmitted to our force-measuring instrument. However, when the seaweed becomes fully extended in the direction the water is moving, it is jerked to a halt, and we suddenly measure a big force on our transducer. After that the alga is no longer moving *with* the water, but rather the water is flowing past the alga. When the water moves relative to the alga, we measure hydrodynamic forces on it that depend on the velocity and acceleration of the water.

Think about holding a dog on a long leash to get a feel for the forces on the holdfast of an alga in waves. If the dog starts to run, you feel no

35 Flexible emerald-green blades of surfgrass are whiplashed back and
forth as water washes past them, to and fro.

36 The leathery brown blades of a seaweed are attached to the shore by a stemlike stipe and an elaborate, branching holdfast glued firmly to the rock.

force on the leash until the dog travels far enough to pull the leash taut. Then suddenly you feel a large force on the leash as it yanks on the dog and stops him in his tracks. The big force you feel at the instant when the dog reaches the end of his rope is due to the rapid deceleration of the mass of the dog. You have to hold on tighter if you've got a Great Dane on the end of your leash than if you have a dachshund. In the same way, massive kelp yank on their tethers with greater force than do skinny, light surfgrass blades. If your dog is running when the leash jerks him to a halt, you feel a much bigger tug on the tether than if he is walking at the instant when he pulls the leash taut. In the same way, the holdfast of an alga is pulled much harder if the alga is moving along with rapidly flowing water at the instant when it reaches the end of its rope than if it is traveling with slowly moving water. Remember that water in a wave speeds up, slows down, and then flows back the other way. If a seaweed is long enough, it may not get jerked to a halt until the water in the wave is slowing down. Sometimes the water starts moving back in the opposite direction before a very long kelp gets fully strung out, so there is no jerk on its holdfast at all. We were at first surprised when we discovered that the forces on the holdfasts of seaweeds three meters long were no greater than the forces on those only one meter in length sitting right next to them on the same rock. Although rigid creatures experience larger hydrodynamic forces as they grow, getting longer can sometimes lead to a reduction in the forces felt by flexible kelp in waves.

Hanging On

Many of the plants and animals covering coastal rocks stay put by gluing themselves to the shore. Barnacles stick their shell houses to rocks with an incredibly strong cement. Likewise, the holdfasts of algae are glued tightly in place (plate 36). In some cases, if you try to pull a seaweed off the shore, its holdfast hangs on so hard that the alga tears or the underlying rock breaks before the adhesive gives way. Mussels make tough *byssal threads,* guy wires that anchor them to the rocks and each other (plate 37). They lay down more threads during stormy seasons, tethering themselves more tightly when the waves are bigger and the hydrodynamic forces are fiercer. Unlike the epoxy we buy at the hardware store, the amazing biological glues that stick the ends of byssal threads, the bottoms of barnacles, and the holdfasts of seaweeds to the shore adhere to wet surfaces.

In contrast to well-connected animals like barnacles and mussels, the soft colonial sea squirts that inhabit protected microhabitats on the shore can come unglued easily. These animals form flat, fleshy colonies pasted to the substratum, enlarging quickly and overgrowing their neighbors as they compete for precious living space (plate 38). If the water flowing across a sea-squirt colony lifts up a flap at its edge, the colony can be peeled off the bottom like a Post-it note. Its soft, squishy tissues are protected from ripping because its adhesive is so weak. A detached sea-squirt colony rafts in the currents and can stick to a new surface if the currents deposit it someplace else.

Not all the animals on a wave-swept shore are glued in place. For example, intertidal crabs can run around in the air at low tide and in the water at high tide (plate 39). You can identify with a crab moving in a tide pool if you imagine yourself trying to run in a swimming pool. When you run, you propel yourself along by pushing on the ground. When you try to run in a pool, it is hard to keep your feet on the bottom because your weight (which is what keeps your shoes on the track when you run on land) is buoyed up by the water. How do you run underwater? You

37 The golden byssal threads that a mussel secretes to tether itself to the
shore are tough and resilient. They act like shock absorbers, stretching a bit
to accommodate the load each time a wave yanks the mussel.

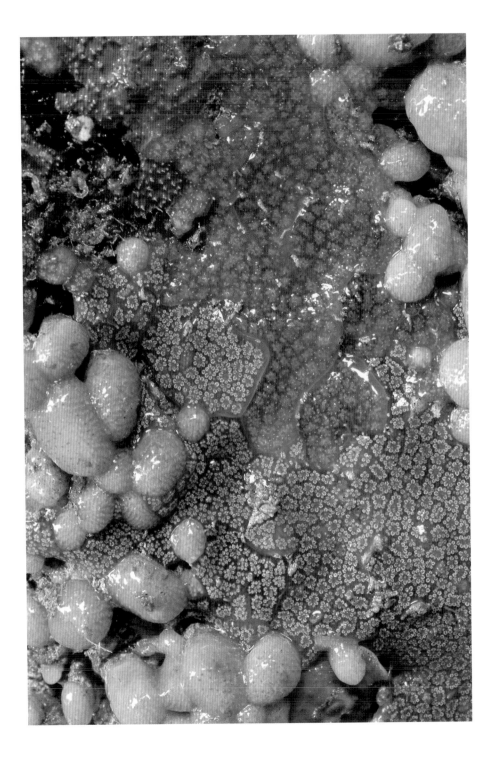

38 Fleshy brown colonies of sea squirts overgrow each other as they compete for space on a rock. These soft slabs can easily be peeled off the bottom, but they can reattach themselves. Each pale petal of the flowerlike clusters visible in the colonies is an individual animal. The white blobs are also seasquirt colonies, and the red patches are sponges *(opposite)*.

39 Intertidal crabs are agile runners. They can avoid being swept away by waves if they find shelter in crevices like this *(above)*.

bound along like an astronaut experiencing low gravity on the moon, pushing off the bottom and then gliding through the water until you sink enough to touch the bottom and push off again. A crab in a tide pool also bounds along, pushing off the rock with one or two feet and then gliding through the water. In contrast, when a crab walks in air, it uses a much more regular gait and always has some of its feet on the ground. As you might imagine, a crab bounding through the water can easily be washed away by a wave. Even a crab just sitting still on the bottom, held in place by its own weight, is in danger of being flipped over or swept away by flowing water. Therefore, crabs often retreat to protected microhabitats or wedge themselves into cracks in the rocks when waves are crashing onto the shore. Water movement can limit foraging forays by crabs.

Other mobile animals at wave-swept sites stick tightly to the substratum while they walk, but they creep along much more slowly than the fleet-footed crabs. For example, some legged animals like pycnogonids (sea spiders, plate 40) and isopods (marine relatives of pill bugs, plate 41) crawl around on plants and animals glued to the shore, hanging on with hooks every step of the way. Starfish also cling to the shore as they move around, walking on hundreds of small, wormlike tube feet (plate 42). Each tube foot has a suction cup on the end and can attach to and detach from the bottom. While some tube feet let go and take steps forward, others are firmly suckered in place, anchoring the animal to the shore. The big, colorful starfish living on our stretch of coastline also use their tube feet to pull open the shells of mussels, their favorite prey (plate 43).

Snails and limpets (which are snails with cone-shaped shells) also stick to surfaces while they slowly plow across them. A snail crawls by passing waves of motion along its single, muscular foot (plate 44). You can see how this works if you coax a garden snail to crawl on a piece of glass and

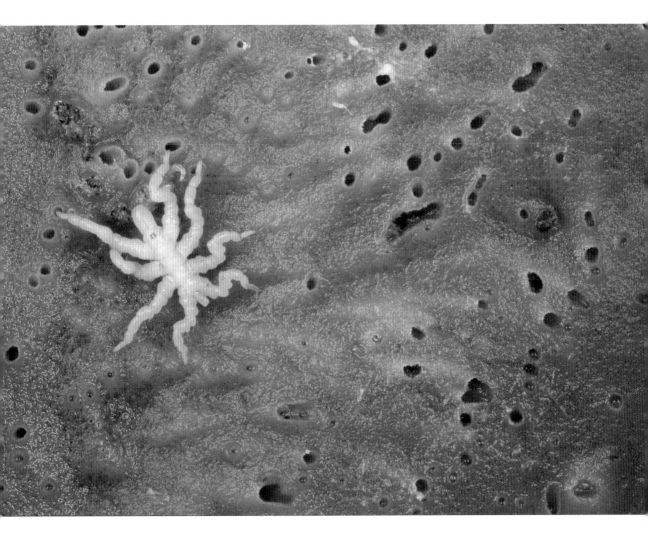

40 A flat, leggy pycnogonid keeps a low profile on a sponge in a protected corner of a cave. It clings to the sponge as it walks, using hooks at the ends of its feet. Pycnogonids are distant relatives of scorpions, spiders, and ticks.

41 A shiny, segmented isopod wedges itself into a clump of goose barnacles, hanging onto them tightly with its many legs. The pill bugs you find under logs in your garden are also isopods, but don't be misled by their common name. Like the isopod in this photograph, they are not insects but rather are crustaceans, like crabs *(above)*.

42 A starfish slowly crawls along a rock wall, hanging there and pulling itself along with its many tiny tube feet. A few of these tube feet can be seen just above the arm on the left side of the animal *(opposite)*.

43 A starfish eating a mussel. The pearllike jewels covering the starfish are the visible tips of little bonelike structures, called ossicles, embedded in the skin. The starfish drapes itself over the mussel and then locks its ossicles in place to form a rigid scaffolding, custom-fit to that mussel. The predator anchors the tips of its arms to the rock with some of its tube feet and attaches the suckers at the ends of other tube feet to each shell of the mussel. The starfish then contracts those tube feet, pulling the shells apart just a crack, while the mussel tries to resist. While this tug-of-war is going on between the mussel trying to pull its shells shut and the starfish trying to pull them apart, the starfish everts its stomach out through its mouth and squeezes it into the narrow gap between the mussel shells. Thus, the starfish starts to digest the mussel while the hapless prey is still fighting to pull its shells closed *(opposite)*.

44 A snail crawls along the bottom of a tide pool on a single muscular foot, which on this individual is highlighted by a pretty orange border. The mucus layer that the snail lays down below its foot has special mechanical properties that let it serve as both a lubricant and a glue, permitting the foot to slip along the substratum as the animal moves, but also sticking the snail to the rock *(above)*.

then watch the bottom of its foot through the glass as the animal inches along. You will also see a trail of slime left on the glass because the snail lays down a carpet of mucus under its foot as it goes. If a snail is to stick to the shore, its mucus must act like a glue; but if the animal is to crawl, its mucus must serve as a lubricant that lets the moving waves along the foot slip easily over the rock. How can one slimy substance serve these two very different functions? Mucus can be both a glue and a lubricant because it exhibits *shear-thinning* behavior. A familiar household product with similar properties is nondrip paint. When you spread the paint on a wall, you shear it (see diagram 6) as you stroke your paintbrush across it, and it flows easily, like a liquid. However, as soon as you stop spreading it, the paint sets up like a solid and does not run down the wall. In the same way, the mucus under the stationary parts of a snail's foot sets up like a solid glue, whereas the mucus under the portions of the foot that are sliding forward is sheared and flows like a liquid. That way only the parts of the foot that are moving are lubricated, while the rest of the foot is glued in place. If the surf is very rough, a snail or a limpet can hang onto the shore more tightly if it stops walking and clamps its shell down onto the rock. Thus, crashing waves can limit when and where mobile animals like snails or crabs can be out and about in search of food or mates.

Stress

What do hydrodynamic forces do to plants and animals on the shore? When a man-made or a living structure is subjected to a force, the materials from which the structure is built must bear that mechanical force. In a plant or animal, those materials are the tissues of its body and the shells and glues it secretes. We can figure out just how hard those materials are

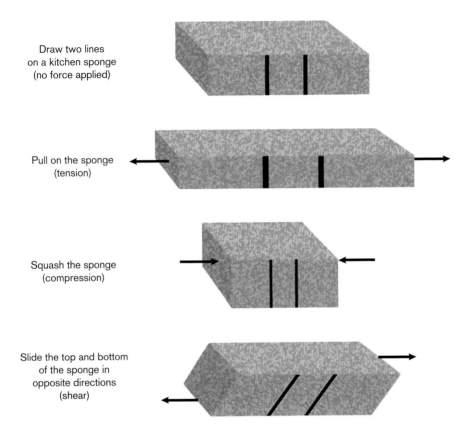

Draw two lines
on a kitchen sponge
(no force applied)

Pull on the sponge
(tension)

Squash the sponge
(compression)

Slide the top and bottom
of the sponge in
opposite directions
(shear)

DIAGRAM 6 If you pull, push, or shear a kitchen sponge, you can see how materials deform when subjected to tension, compression, or shear.

pushed, pulled, or sheared by the hydrodynamic forces on a living thing by analyzing it in the same way that an engineer figures out how mechanical forces are distributed in the struts of a building or bridge.

Think about hanging an apple on a thread. The apple's weight is a force that pulls on the thread. Now imagine hanging another apple on a thick rope instead. The weight of the second apple is supported by the many strands that make up the rope, so the pull on each individual fiber in the rope is less than the pull on the single, thin thread supporting the first apple. The mechanical *stress* in the thread or in the rope is the force exerted by the weight of the apple divided by the cross-sectional area of the thread or the rope being pulled by that force. (The cross-sectional area of the thread or the rope is the area you see if you cut it and look at the cut surface end-on.) The stress in the thread is much bigger than the stress in the rope. If we know the shape and size of a part of a living thing, like an algal stipe, and if we know the magnitudes and directions of the forces imposed on it, we can figure out how big the stresses are in the tissues of that structure.

You can use a kitchen sponge to visualize how the materials in a structure are deformed when you apply forces to the structure in different ways (see diagram 6). Draw two lines on the side of the sponge, across its thickness. If you pull on the ends of the sponge, the lines move farther apart. You are deforming the sponge in *tension,* and its material is being stretched, like the rope from which our apple was hanging or the byssal threads tethering a mussel to the shore (plate 37). Now push on the ends of the sponge. If you are careful to keep the sponge from bending or kinking, you see that the lines you drew on the side of the sponge move closer together. You are deforming the sponge in *compression,* and its material is being squashed, just as the tissue of a sea squirt colony (plate 38) is mashed when you stand on it. Now lay the sponge on the palm of one hand and place your other hand on the top of the sponge. If you slide your hands in opposite directions, the lines you have drawn on the side of the sponge tilt over at an angle. You are *shearing* the material in the sponge, trying to make one layer slide sideways relative to the layer just below it, the way

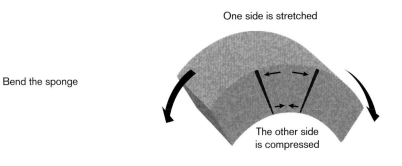

One side is stretched

Bend the sponge

The other side
is compressed

DIAGRAM 7   If you bend a sponge, one side of the sponge is stretched, while the other side is compressed.

that the mucus under the foot of a crawling snail is sheared. For tensile, compressive, and shearing loads, the greater the cross-sectional area of the structure bearing the force, the lower the stress in its tissue or glue.

Many living things, like trees in the wind and algae in currents, are bent by fluid-dynamic forces. Bending is a more complicated way to deform than simple tension, compression, or shear. If you bend your kitchen sponge (see diagram 7), the stripes you drew on the side of the sponge move farther apart on the convex side of the bent sponge (where the material is in tension), but are squashed closer together on the concave side (where the material is under compression). Engineers have developed ways to determine how big the tensile and compressive stresses are within a bending beam, and we use those same calculations to figure out how large the stresses are in plants and animals being bent by waves. It turns out that the longer the beam (or body part), the larger the stresses in it

when it is bent by a force on its end, so the tissues in a tall plant or animal experience higher stresses than do those of a shorter individual of the same width subjected to exactly the same bending force!

For bending beams like algal stipes that are attached to the substratum, the biggest stresses occur near the bottom of the stipe, where it attaches to the rock, *if* the stipe is the same width all along its length, from base to tip. However, tree trunks and algal stipes are often wide near the bottom and tapered toward the top (plate 45). How does the width of a bending beam affect the stresses in its tissues? Remembering the apple hanging on the thread versus on the rope, we would expect the stresses to be lower where a beam is wider. This is true, but in a bending beam the effect of width is much more dramatic than it is in a thread being pulled. If you double the diameter of a bending stipe, you reduce the stress by a factor of eight, and if you triple its diameter, you reduce the stress by a factor of twenty-seven. Therefore, if a seaweed thickens its stipe just a little bit, it can reduce the stress in the tissues of its stipe by an enormous amount. Conversely, if a chomping sea urchin or a grazing limpet eats away part of a seaweed (plate 46), the local stress at the skinny, munched portion of the alga can be much higher than in the surrounding tissues. During storms, seaweeds often break at these high-stress grazed spots, so the algae are pruned to a smaller size by the waves rather than being completely ripped from the shore.

45 A sea palm (an alga that looks like a little palm tree) that was ripped from the rocks, complete with its holdfast, and washed ashore on a sandy beach. When the alga was attached to the shore, its tapered stipe stood upright, holding its many straplike blades up above the seaward face of a rock subjected to crashing waves. The rubbery stipe was bent over each time the sea palm was hit by a wave but bounced back upright after the wave passed—that is, until a very big wave tore away its toehold on the rock *(opposite)*.

46 A grazing limpet has rasped away a chunk of this seaweed with its filelike tongue. The alga is likely to break at this thin spot when awash in the wild waves of winter, but such pruning can save the remaining stump of the seaweed from being swept away during storms. The pruned seaweed survives and can continue to grow.

Strain

What do stresses do to the tissues of animals and plants? Just as the material in the kitchen sponge deforms when you push or pull on it, so do the tissues of the animals and plants on a rocky shore when moving water pulls, squishes, shears, and bends them. The local deformation of a material is called *strain*. Some materials, like rubber, chewing gum, and the tissues of floppy algae or squishy sea anemones, deform a lot when subjected to forces. In other words, deformable materials undergo a lot of strain when they are stressed. In contrast, other materials like stiff concrete, steel, and the rigid shells of barnacles, snails, and mussels, strain very little when stressed.

How much a whole structure deforms when exposed to a hydrodynamic force depends both on its shape (which determines how big the stresses are in its tissues) and on the stiffness of its tissues (which determines how much they strain in response to those stresses). You can demonstrate this for yourself at the hardware store. Try bending different pieces of wooden doweling. Even though they are all made of wood, it is much easier to bend a long dowel than to bend a short one of the same diameter, and it is much harder to bend a thick one than a slender one. Now try bending rods made of steel, a stiffer material than wood. It's harder to bend a steel rod than a wooden dowel of the same dimensions. With this in mind, it makes sense that long, slender algae made of stretchy tissues deform easily when water washes over them, while short barnacle shells made of stiff calcium carbonate sit there rigidly resisting the flow.

Structures made of stretchy materials, like seaweed stipes and mussel byssal threads, can act like shock absorbers. By stretching when yanked by a wave, these extensible structures absorb a lot of mechanical energy, just like the elastic cord that eases the fall of a bungee jumper. Sudden high forces on an alga or mussel are damped by deformation of the stipe or byssal threads, so their attachments to the rock are protected from big jolts when waves hit. Some stretchy biomaterials, including those found in seaweed stipes, are very resilient and bounce back to their unstretched shape right away, like rubber bands.

Breaking and Washing Away

Animals and plants on a wave-swept shore are ripped or broken if the stresses imposed on their tissues by rapidly moving water are greater than the strength of those tissues. *Strength* is the stress required to break a particular material or tissue. However, in order to understand what makes some biomaterials more breakable than others, we also need to think about how much energy is required to drive a crack through the material while it is fracturing in two. Glass is stronger than leather, but if you throw a vase and a shoe at the wall, the vase is more likely to break than the shoe because leather is tough and glass is brittle. A tough material absorbs and dissipates a lot of the energy that you put into it as you try to tear or smash it apart. In contrast, a brittle material transmits that energy right to the crack that is moving across the material as it breaks, and the material fractures catastrophically.

Scientists have studied some of the biomaterials used by wave-swept animals and plants to discover what makes them hard to break and to learn tricks from nature for the design of tough man-made materials. Consider the material composing the shell of a sea snail. Shell is primarily calcium carbonate, the same substance that makes up a piece of chalk. However, shell is tough, and chalk is brittle. The secret to the toughness of shell lies in its architecture at the microscopic scale: it is a complex composite material. Mother of pearl, the shiny opalescent layer lining the inner surface of a shell, is made up of a beautiful brickwork of tiny tiles of crystalline calcium carbonate, each only a few thousandths of a millimeter thick, while the outer layers of a shell are composed of fine needlelike crystals laid crisscross, layer upon layer, like plywood. A pliant matrix is woven through these tiles and needles like the mortar between bricks. If one tile or needle in this complex composite material breaks, the crack that propagated through the brittle crystal stops when it hits the mortar. The crack gets diverted along the surfaces between the microscopic bricks rather than continuing on a nice straight path across the next neighboring piece of calcium carbonate. The fracture propagates here, there, and everywhere, taking a tortuous route through the complex array of chalk tiles and the glue between them. It takes lots of mechanical energy to drive that crack all the way through the whole piece of shell to break it. Our own bones, though made mainly of calcium phosphate rather than calcium carbonate, are toughened in much the same way.

Stretchy materials like your skin, the sea anemone's body wall, and kelp stipes are also complex composite materials. Rather than being made up of brittle crystals and tough mortar, like shell, they are woven from strong, hard-to-stretch microscopic fibers embedded in a much softer, easily deformed matrix. The fibers in sea anemones, and in us, are made of proteins, as are the silvery threads that attach chicken meat to the bones. The fibers in seaweeds, seagrasses, and sea squirts are made of carbohydrates, as are the strings in a stalk of celery. The way those fibers are arranged and cross-linked together (in tight cables or loose mesh), along with the stretchiness and springiness of the matrix between them, determine just how deformable, resilient, and tough the whole composite tissue is. In some fleshy animals, like sponges and starfish, tiny bits of hard grit are embedded in the soft tissues. In many types of sponges, these bits of stiff material are made of silicone dioxide (glass) and are called *spicules*. In starfish, they are made of calcium carbonate and called *ossicles*. They stiffen the tissues in which they are embedded in the same way that microscopic carbon particles are used as filler to stiffen the rubber in car tires.

Sometimes streamlined shapes and tough biomaterials are not enough to prevent breakage on a wave-swept shore. Waves hurl logs onto the rocks and roll boulders across the shore. These projectiles and bulldozers mow down the living things in their path, cutting swaths of destruction through the aggregations of animals and plants carpeting the rocks (plate 47). In spite of their grim appearance, patches of bare rock (plate 48) scraped clean of living things by wave-tossed battering rams play an important role in maintaining the rich diversity of species living on the shore. Certain species of mussels, seaweeds, and barnacles are more effective at competing for space on the rocks than other species, overgrowing or crowding out their neighbors. If left unchecked by predators or waves, monocultures of such dominant space-holding species would cover the rocks. However, bald spots created by wave-tossed logs are colonized by a variety of plants and animals. It takes several years for the most successful competitors to take over that space, so in the meantime other species can thrive there. Thus, because waves and the projectiles they hurl at the shore rip living things off the rocks, a diverse patchwork quilt of many species can develop on the shore. As the threadbare spots in this quilt are mended by new growth, its texture is made far richer.

47 The holdfasts and stumps of stipes of sea palms (plate 45) that have been
mowed down by a wave-borne log scraping its way across the rocks.

48 A wave-tossed log crashing into the shore like a battering ram created the stark, bare patch of pale rock in the middle of the rich carpet of life on the shore.

Diverse Architectures Work

The body forms of those individuals in a population that successfully compete with their neighbors for the resources they need to grow, and that survive the surf and escape predation long enough to reproduce, are represented in the next generation of that species to populate the coast. Thus, generation after generation, natural selection gradually weeds out the bad body forms, leaving behind those just good enough to hang on, grow, and produce offspring. This process of natural selection has, over the ages, sculpted the shapes of the plants and animals that survive on the shore.

As we step back and once again survey the smorgasbord of living forms clinging to the rocks of a wave-swept shore, we now see with an engineer's eye which forms are subjected to large hydrodynamic forces and which ones are not. We can recognize the body shapes that experience high tissue stresses, and we can predict the consequences of pulling or pushing on the different types of biomaterials from which they are built. Tall, slim algae made of stretchy, resilient material sway back and forth in the waves, whereas short, wide barnacles with their stiff, calcified shells sit unmoved as the water crashes onto them. By studying the mechanics of these plants and animals, we come to understand how such diverse structures can all function in a wave-swept environment.

5

TRANSPORT

Motion and Mixing

How do animals glued to one spot on the shore forage for food? How do seaweeds, which do not have root systems, get their nutrients? How do the plants and animals stuck to the rocks breed if they can't move around to find mates? How do they migrate to new habitats? The water swirling around them performs these life-sustaining functions. Even though the living things clinging to the shore are at risk of being swept away by waves and currents, they also depend on that moving water to bring them many of the things they need to live.

Dissolved substances in the ocean, such as oxygen, nutrients, and odors,[1] are carried in the water as it moves. Likewise, objects as tiny as microscopic bacteria and as big as logs are moved from place to place by flowing water. Currents and tidal flows transport water-borne materials long distances along the coast or between the shore and the open ocean. Close to the shore, the back-and-forth motion of the water in waves slowly sloshes materials towards the beach.

In the roiling sea, eddies large and small swirl around in confusion.

---

1. *Odors* are made up of molecules carried in a fluid, such as water or air.

49 As waves move towards the shore, the water in them sloshes back and forth and turbulent eddies spin in all directions *(opposite)*.

Because the ocean is turbulent, materials carried in the water get mixed up and dispersed, just like suds and socks do in the churning water in a washing machine. Tiny vortices stir up the water flowing across the surface of a seaweed, while huge gyres mix water across many kilometers. To get an idea of how substances are dispersed in the water, think about how smoke from a chimney behaves. A smoke plume drifts off in the direction the wind is blowing; in the same way, materials suspended or dissolved in the ocean travel in its currents. As a smoke plume is carried along in the turbulent wind, eddies stir the smoke into the surrounding air, so the plume becomes wider and more diffuse as it is carried away from the chimney. Similarly, materials carried in the water are dispersed and diluted by the turbulent eddies in the sea.

Dinner Delivery

The demise of one animal in crashing waves can provide the dinner for another. Mussels ripped from the rocks by the surf are captured by the large green sea anemones lurking in the channels below them (plate 50). Because they are short and wide, these anemones do not bend over in the flow (as explained in chapter 4), so their tentacle-encircled faces point upward, at the ready to trap prey swept onto them by the surf. The waves are their waiters. Because sea anemones swallow their prey whole and disgorge the shells after digesting their meal, we can easily survey the diets of these opportunistic predators. The big green anemones eat animals that live above them on the shore, so anemones below mussel beds eat mainly mussels, and those below sea urchins often feast on that spiny fare. Every so often, anemones living below gull rookeries have an egg for breakfast! In spite of their lovely appearance, the big green sea anemones that carpet the bottoms of surge channels are the garbage cans of the intertidal.

Predators like sea anemones must be able to hang on to the food that is

50 Sea anemones sit in a pool below a mussel bed, waiting for the waves to
wash prey into their grasp. A crown of tentacles surrounds the face (called
the *oral disk*) of a sea anemone. The mouth of the sea anemone is located at
the center of its oral disk. A sea anemone has no anus, so after it digests its
meal, it spits the wastes back out of its mouth.

tossed onto them by the crashing surf. When a mussel or sea urchin lands on a green sea anemone, the tentacles curl over the prey, sticking to it with a fur of microscopic barbed threads, fired like miniature harpoons from special cells covering the tentacles. The waves can't snatch away the prey because the anemone hugs its quarry tightly with these adhesive tentacles as its face folds over the food and its lips encircle the meal.

Sea urchins are herbivores that eat algae. Urchins can take bites out of seaweeds, and they can graze on food that they scrape off the rocks. Another important component of urchin diets can be drifting pieces of broken seaweed *(drift algae)* that are carried in the water flowing past them. Sea urchins have tube feet, like those that their starfish relatives use to stick themselves to the rocks (plate 42). Urchins hang on to drift algae with their tube feet and use their spines to move the captured salad toward their mouths.

Filtering Food from the Flow

A myriad of microscopic bacteria, single-celled algae, protozoans, and tiny animals live suspended in the waters of the ocean; they are called *plankton,* which means *drifters*. The water is also laden with tiny pieces of organic matter, including bits of decomposing plants and animals, mucus and fecal pellets shed by animals, and bacteria-covered grains of suspended sediment. Many animals on the shore make their living by filtering these miniscule creatures or particles from the water; this form of food-gathering is termed *suspension feeding*. Currents in the habitat carry plankton-laden, particle-rich water into the vicinity of the suspension feeders on the shore and in some cases even drive the water through their food-catching filters.

Goose barnacles (plate 51) feed on tiny animals that they strain out of the water. Barnacles are crustaceans, like shrimp and crabs, and have a set of bristly legs curled up inside their armor of calcareous plates. After a wave breaks on the shore, the water washes down the faces of the rocks, carrying with it *zooplankton* (planktonic animals) and little creatures washed loose from the substratum. A goose barnacle spreads its legs out like a fan into this backwash, filtering food from the water flowing between its bristled appendages (plate 52). If you watch goose barnacles feeding in the surf, you will discover that each animal is oriented so that its fan-like row of legs is spread at right angles to the direction of the backwash over that bit of rock. This orientation of the food-catching net, broadside to the flow, allows the barnacle to intercept the widest possible swath of water washing across the rock. If you stand back and look at the patterns of how goose barnacles are aimed, you'll see a living map of the directions in which water flows across them on its way back to the sea.

Animals like goose barnacles that depend on ocean currents and waves to drive water through their food-catching nets are called *passive suspension feeders*. In contrast, their relatives the acorn barnacles (plate 53) are *active suspension feeders* that drive the water through their particle-catching devices themselves. Acorn barnacles can filter food particles from still or slowly moving water by rhythmically sweeping their bristly legs through the water. However, if the water flow in their habitat speeds up faster than their legs can move, acorn barnacles switch to the passive mode of suspension feeding used by goose barnacles, simply holding their fan of particle-catching appendages broadside to the water flow. However, while goose barnacles only strain their food out of the wave backwash flowing down the rocks, acorn barnacles actively aim their cup-shaped filtering fans to face upstream. In waves, acorn barnacles flip their filters around every few seconds, keeping pace with the back-and-forth movement of the water across them.

Other active suspension feeders, like the sponges and sea squirts plastered

51 Goose barnacles at low tide, with their legs tucked away behind their body armor *(above)*.

52 Feeding goose barnacles stick their legs out into the water washing off the rocks. The fan of legs of each barnacle looks like an upside-down black spider poking out of the barnacle's shell, or perhaps like a sinister hand with many dark fingers, cupped to catch the water running down the rock. A goose barnacle catches tiny prey from the water flowing through the filter formed by its fan of hairy legs *(opposite)*.

53 White acorn barnacles, red sponges, and tan sea-squirt colonies on the walls of a cave are all suspension feeders that actively drive water through their filters. The large holes in the sponge in the lower left are the nozzles through which food-depleted water flows rapidly out of the sponge (see diagram 8). Dead barnacles near the little black and pink starfish are being overgrown by red sponges.

on the walls of our sea cave (plate 53), or the mussels clinging to wave-exposed rock faces alongside the goose barnacles (plate 54), pump water through food-capturing plumbing systems (see diagram 8). In sponges, an elaborate network of branching pipes carries water through the sponge, into and out of special chambers lined with microscopic whiplike structures called *flagella,* which actively wave around to pump the water. Food particles carried in the water are trapped on the walls of the pipes and on comblike collars surrounding each flagellum. Individual sponge cells engulf and digest these particles in a manner similar to the way that an amoeba engulfs its prey.

A sea squirt draws water through a siphon into a large, basketlike bag (called a *pharynx*) that is perforated by many slits lined with *cilia* (microscopic whips that are like flagella, but shorter). The beating of these cilia drives water through the slits, moving it from the inside of the pharynx into a space surrounding the pharynx, from which it exits the animal through a narrow nozzle. Food particles are trapped on a net of mucus lining the pharynx. The mucus net is moved along the pharynx like a conveyor belt, carrying the trapped particles to the sea squirt's mouth; there they are consumed, net and all. Some species of sea squirts can bud off replicates of themselves to create colonies in which many individuals are embedded together in a thick, fleshy sheet plastered on the rock (plate 38). In such a colony, the individual sea squirts each have a food-filtering pharynx, but groups of neighboring individuals funnel all their processed water into a common exit chimney.

Mussels have large, filamentous gills that hang like grass skirts in the water-filled spaces inside their shells. The gills are covered by bands of flapping cilia, some of which are aimed so that they drive water through the gaps between gill filaments. Single-celled algae carried in the water are caught on the gills, and special rows of cilia sweep them towards the

54 A feeding mussel sits among goose barnacles with its shells gaping open as it pumps water through its particle-capturing gills (see diagram 8).

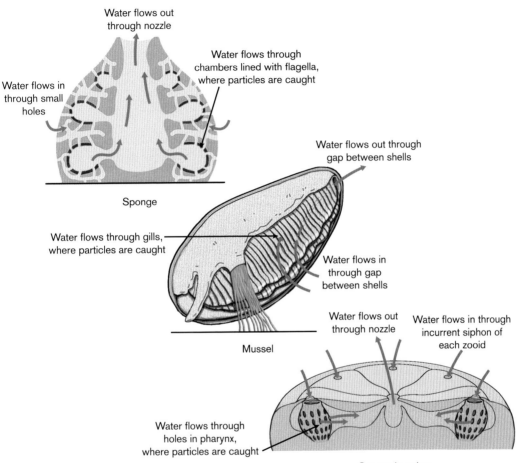

Water flows out
through nozzle

Water flows through
chambers lined with flagella,
where particles are caught

Water flows in
through small
holes

Sponge

Water flows out through
gap between shells

Water flows through gills,
where particles are caught

Water flows in
through gap
between shells

Mussel

Water flows out
through nozzle

Water flows in through
incurrent siphon of
each zooid

Water flows through
holes in pharynx,
where particles are caught

Sea-squirt colony

DIAGRAM 8 Cutaway views of some animals that pump water through themselves.
Blue arrows show where the water moves, and black arrows indicate the places in the
animals where food particles are filtered from the water pumped through the animals.
Sponge. Water flows in through small holes at the side, into chambers lined with flagella
where food particles are caught, and out through a nozzle. *Mussel*. Water flows in
through the gap between the mussel's shells, through gills, where particles are caught,
and out again through the gap between the shells. *Sea-squirt colony*. Water flows in
through the incurrent siphon of each animal, through holes in the pharynx, where
food particles are caught, and out through an exit nozzle shared by a group of animals.

mussel's mouth. When a mussel is feeding, its two shells gape apart slightly (plate 54) so that unfiltered, food-rich water can flow into the spaces around the gills, and filtered water can be expelled. If you gently drop food coloring or milk into the water near a feeding mussel, you can see that water enters the cavity between the shells in one area and exits from the animal more rapidly at a different place.

Sponges, sea squirts, and mussels expel the water that they have filtered through narrow exit nozzles. Remember how water accelerates through the small exit window of our cave (diagram 3) or the end of a garden hose that you've partially covered with your finger. In the same way, the food-depleted waste water from a mussel, sponge, or sea squirt speeds up as it is forced through a narrow exit opening. This waste water flowing rapidly out of a nozzle not only is propelled away from the water-intake ports of the animal, but also is squirted away from the substratum, up into the rapid flow that can carry the used water away from the neighborhood. Thus, by using a nozzle to squirt food-depleted water rapidly away from its body, a suspension feeder reduces its chances of sucking in and reprocessing the same old water over and over.

Because they do their own pumping, active suspension feeders are able to capture particles in still water. These impressive filtering systems can, however, quickly deplete the water around them of food. Even active suspension feeders depend on ocean waves and currents to replace the old, depleted water in their neighborhood with new, food-rich water.

Solutions for Life

Many molecules essential for life are dissolved in the seawater washing over the plants and animals on the shore. Algae don't have roots, but rather take up from the water around them the dissolved nutrients they need for growth and the carbon-containing compounds that are the raw materials for photosynthesis. When submerged, intertidal animals get their oxygen from the surrounding water. Algae produce oxygen during photosynthesis, but in the dark of night they too pick up oxygen for respiration from the water around them.

When an animal or plant sitting in a still pool takes up dissolved substances, it depletes the water right around its body of those molecules. In contrast, if the water around the creature or alga is flowing, this depleted water is continuously washed away and replaced with new, fresh supplies. When the boundary layer was introduced in chapter 3, it was discussed in terms of the flow along the substratum. Boundary layers also form along the surfaces of plants and animals in flowing water or air. Therefore, even in moving water, a halo of depleted water builds up around the individual because water moves slowly in the boundary layer along its body surfaces. Remember that the boundary layer, and thus the halo, is thinner in wavy flow than in a current flowing steadily in one direction. Also remember that the faster the flow, the thinner the halo. Furthermore, if the water movement is very turbulent, then the swirling eddies can sweep away stale water along an individual's surface and stir undepleted water down into the boundary layer. Thus, when we measure the rate at which an animal takes up oxygen for respiration or an alga takes in bicarbonate for photosynthesis, we find that the faster or more turbulent the water flow, the greater the uptake rate.

If the flow gets going fast enough and the boundary layer becomes thin enough, the supply of nutrient or oxygen molecules in the water next to a plant's or animal's surface no longer limits the rate at which it can take in the molecules. Instead, factors like the rate at which its cells can use the molecules set the limit on how quickly those substances move into its body. For some types of kelp, the water velocity at which the supply of molecules no longer limits the rate of photosynthesis (this water speed is called the *saturation velocity*) is about one meter per second. That means that at slack tide on calm days when the waves are small, photosynthesis by these kelp can be limited by the rate at which raw materials are supplied by the water. However, when tidal currents are flowing and large waves wash across the kelp, the water typically moves faster than the saturation velocity.

Inhabitants of the shore use various tricks to enhance the rates at which they can supply themselves with dissolved materials. Many animals, like the sea squirts and mussels described above, actively pump water through their gills, which take up oxygen as well as capture food particles. Many seaweeds flutter in the flow like flags in the wind, and those with ruffled blades tend to flap around more violently than do those with flat blades. This fluttering stirs up the water around the algae and increases uptake rates when the currents across the seaweeds are moving more slowly than saturation velocity.

Waste Disposal

Marine plants and animals release dissolved wastes into the water around them. For example, most marine animals excrete their nitrogenous wastes in the form of nasty solutions of toxic ammonia. Animals also dispose of solid wastes, such as the inedible particles dropped by mussels, the undigested shells disgorged by sea anemones, or the feces pooped out of the guts of many species. Such dissolved and solid wastes can foul the still water of a stagnant pool and can build up in the boundary layers along the surfaces of animals and plants.

Animals and algae stuck in one spot on the shore rely on the water motion around them for waste disposal. The same factors that enhance the supply of dissolved materials or solid food to these inhabitants of the shore also improve the removal of waste products. Rapid, turbulent flow sweeps away fouled water, and the fluttering and flailing of flexible plants and animals in the waves enhances this process by stirring the water and sweeping the substratum. Animals that pump water through themselves, like sponges, sea squirts, and mussels, release their ammonia-ridden urine and their feces into the food-depleted waste water that is squirted away from their bodies through narrow nozzles, as described above.

Deposits, Debris, and Destruction

Moving water can transport objects, like sand grains, pebbles, and beer bottles, from one place to another. The speed of the water over a patch of bottom determines whether objects sitting on the bottom will be moved. In slow flow, only the lightest particles, like fine silt and small bits of biological detritus, are washed away. As the velocity picks up, bigger, heavier things, like sand grains, begin to roll and bounce along the bottom. When the flow is faster still, sand can be swept up into suspension in the water and carried away. At very exposed sites, pebbles and cobbles are moved around by the water and can be flung by waves. In fierce storms, even boulders can be picked up and hurled against the shore by the pounding surf.

When moving water slows down, objects in suspension sink back to the bottom. Tiny, light particles, such as microscopic pieces of detritus, drop out at slower water velocities than do larger, heavier bodies, such as sand grains or pebbles. In very slowly moving water, most of the suspended material settles out of the water and lands on the substratum below. Therefore, animals and plants living in still water can become covered with silt and detritus

55 The gentle waves of summer deposit sand on the shore, building up the beach and burying sea anemones. The anemones can survive this burial if oxygenated water percolates past them through the porous sand. The wild waves of winter will once again sweep the sand away, uncovering the anemones and chewing away the beach. The seasons and the sands come and go, year after year *(above)*.

56 Loose cobblestones tumbled around in a pool by waves are scraped clean of animals and plants *(opposite)*.

unless they have ways of cleaning their surfaces. Some of the creatures growing on our stretch of shoreline are buried by sand in the summer, when the waves are small and the flow is slow (plate 55). They are exhumed again in the winter, when big storm waves wash the sand away.

Large objects carried by moving water can wreak havoc on the inhabitants of the shore. The damage done by floating logs flung onto the rocks was described in chapter 4 (plates 47 and 48). But even marble-sized pebbles thrown by waves can crush barnacles and chip limpet shells (plate 68). Cobbles rolling around in the surf can also smash animals and algae clinging to the rocks or can scrape and wear them away over time (plate 56).

Sex in the Sea

It is hard for boy to meet girl (or for a hermaphrodite to run into a mate) when everybody is glued in place on the shore. Nonetheless, attached marine animals and plants do manage to have sex. Many of them spawn, releasing their sperm and eggs into the surrounding sea, where turbulence stirs the water and mixes them together. Fertilization occurs when sperm and eggs meet in the water—a chancy process at best. Marine animals attached to the bottom can improve their odds of breeding successfully by spawning at the same time as the other members of their species in the neighborhood. If they all build up their stores of eggs or sperm during the same season, they'll be ready to breed at the same time. In some species, odors emitted when one animal spews out its eggs or sperm are spread around the neighborhood by turbulent waves and currents, and these aromas stimulate other members of that species to begin spawning as well. An orgy of swirling eggs and sperm ensues, and the chances of an egg meeting up with sperm become much greater.

You might think that the more turbulence and mixing there is in the water, the better the spawning success. To a degree this is true. However, if the flow is too violent, delicate eggs and sperm can be damaged. In addition, if the water flow across a habitat is too rapid or turbulent, the eggs and sperm can be dispersed and diluted so rapidly that few of them bump into each other. Some species of animals get around this problem by packaging their eggs or sperm in goop. Mucilaginous strings of eggs and sperm are twirled and swirled around together by the water flowing over the spawning animals, and chances for successful fertilization are improved. In other species of attached animals, only the males spawn. These animals, as well as seaweeds and seagrasses, rely on water currents in the environment to carry their sperm to females, who catch the sperm to fertilize their eggs. Some of these mothers brood their developing young, protecting them in or on their bodies, while mothers of some other species lay their fertilized eggs, gluing them in clusters to the rocks.

Although most of the types of animals that live stuck to the sea floor have sex by spawning, barnacles are a notable exception. A barnacle has a penis that is many times longer than its body. The penis is so long that it can reach other barnacles glued to the rock nearby. Since barnacles are hermaphrodites, they can mate with any barnacle of the same species that happens to be living within reach. If you bolt a camera to the rocks to record what barnacles do when the tide is in, you find that they spend a lot of time filtering food with their bristly legs, but they also spend time snuffling around the neighborhood with their penises, perhaps checking out potential mates.

A number of the animals that can move about on the shore, like crabs and some kinds of snails, copulate rather than spawn. Unlike barnacles, they can cruise around looking for mates rather than being stuck with the boy/girl next door. Copulation seems much less chancy than spawning, since sperm are deposited directly into the egg-producing parent rather than tossed into the turbulent waves. However, copulation is still a risky business. If you are an animal that copulates, you must find a suitable mate—a member of the same species that is sexually mature, of the opposite sex (this job is much easier if you are a hermaphrodite, of course!), ripe with eggs or sperm, and willing to breed with you. A potential partner may prefer to eat you than to mate with you, and you may have to compete with other suitors. Since you can only produce a finite number of eggs or sperm, how do you choose which potential mate is likely to be the best father or mother for your offspring? Many animals use chemical signals released into the water to attract and communicate with potential mates. These odor signals are dispersed by the water, like smoke plumes in the wind. Even though animals that copulate do not depend on water currents and waves to transport and mix their eggs and sperm, they may rely on that flow to get their message out.

Travel to New Homes

How can plants or animals stuck to one spot on the shore ever move to greener pastures or colonize distant stretches of coastline? The fertilized eggs of many sedentary marine animals develop into microscopic larvae that swim in the water rather than sit down on the bottom. Although they can swim, some by waving rows of cilia and others by flapping bristly legs, these larvae move too slowly to travel great distances under their own steam. Rather, it is the powerful currents of the ocean that transport these tiny wayfarers to distant shores.

When larvae develop from spawned eggs swirling in the water, escape from egg cases glued to the rocks, or are spewed into the sea by a mother on the shore, they embark on a dangerous voyage. They are dispersed by the same turbulent mixing that spreads and dilutes the wastes released by animals and plants on the shore and the same tides and currents that carry them away. Many larvae are eaten by predators in the plankton or are filtered from the water by suspension feeders, while others simply wash out to sea and never encounter a suitable site to settle. Because the odds of surviving the larval journey are so low, mothers that produce huge numbers of babies have the best chance that at least a few of their children will make it. The larvae of some species (for example, sea squirts, limpets, and some snails) carry fuel supplies provided by mom, while others (like barnacle and mussel larvae) can fend for themselves, catching meals of single-celled algae or other microscopic plankton drifting in the water around them. As larvae are carried by the currents, they develop and mature, eventually becoming capable of metamorphosing from swimming larvae into bottom-living juveniles. When larvae become old enough to undergo metamorphosis, they are called *competent.* Larvae of some species become competent in a few minutes or hours, whereas those of other species can take days, weeks, or months. The longer larvae drift in the ocean, the farther they can travel, but also the greater the risk they face of meeting up with dangerous planktonic predators.

How can tiny larvae, at the mercy of waves and currents, manage to settle down onto the bottom at a patch of shore where the conditions are right for their survival? Those fortunate larvae that happen to be drifting in currents that move near to the coast can ride the waves in to the shore as the water in which they are carried gradually sloshes its way toward the land. As the larvae are carried across the rocks, they are swirled around, up and down, with the passing of each wave. At the same time, turbulent eddies in the roiling flow mix and spread larvae around in the water, much as rice and vegetables are spread around when you stir a pot of soup. Larvae that happen to bump into the bottom can decide to stick if they like it there or to keep swimming if they don't.

What makes a larva accept or abandon a landing site? Larvae of some species reject a landing site if it smells like the animals that prey on them. In contrast, if the place they land tastes or smells right, many larvae are induced to glue themselves down and undergo metamorphosis. The appealing flavor or scent may be from members of the same species already living on that spot of rock—a sure sign that it is a suitable site—or from the prey that the new settlers will eat after they have metamorphosed into bottom-living juveniles. The larvae of some species are induced to sink if they encounter such an aroma wafting in the water; this sinking behavior increases their chances of bumping into the bottom at sites where their future prey are abundant. Certain types of larvae stick to rocks where other larvae of their species have just attached; those that survive until they are old enough to spawn will have potential mates nearby. Even without chemical cues, many larvae collect like particles of sediment in pockets of slowly moving water in crevices and among plants and animals already living on the shore. The spaces within beds of mussels, turfs of algae, and fields of barnacles can provide protected nurseries where delicate larvae can attach themselves and avoid being washed away by the waves. Larvae of many species are picky about where to settle when they first become competent, but as the clock keeps ticking and their time as larvae runs out, they become less fussy and will settle just about anywhere they touch down.

Moving water also spreads the seeds shed by seagrasses and microscopic spores released by seaweeds. In addition, some marine plants have another mode of dispersal. Certain algae and seagrasses are buoyant. If these floating plants are ripped from the shore, they drift along with the currents rather than sink to the bottom. If they dribble spores or seeds as they go, or deposit them where they wash ashore, they can colonize sites farther afield than can spores or seeds released from attached plants. Animals like sea squirts and small crustaceans living on the blades of plants that are torn from the shore can hitch rides on these rafts. Some of these flotillas drift ashore, and the stowaways that can jump ship and hang on to surfaces at that landing site are able to set up shop in a new home. The rafting sea-squirt colony described in chapter 4 can readily reattach to new surfaces if it is torn from its substratum. The larvae of this species carry only enough fuel to swim for a short while (less than an hour) and thus do not travel very far before they settle onto a surface and undergo metamorphosis. It is the floating adult colonies that establish a foothold in distant locations.

Trade-Offs

The swirling sea serves the plants and animals on the shore in many ways, bringing them sustenance, sweeping them clean, and transporting their young. Yet, at the same time, the fury of the surf can tear these living things from the rocks. Animals and plants that live sheltered from rapid flow in the midst of aggregations can be protected from such dangers, but they may be bathed in foul water depleted of food or nutrients by their upstream neighbors. On the other hand, plants and animals that stand tall in the face of the oncoming flow are bathed in rich resources but must be tough enough to withstand the force of the waves. Life is full of trade-offs.

6

STRANDED, HIGH AND DRY

As waves wash over rocks at high tide, they bathe the plants and animals on the shore in the cool water of the ocean. A few hours later, at low tide, ebbing water leaves the sea creatures and algae to face the rigors of terrestrial life. We are well equipped to survive in air, with our waterproof skin and our ability to regulate our own body temperature. In contrast, marine plants and animals without these capabilities can perish if stranded in air too long without the protective buffering of seawater. On sunny days, inhabitants of the intertidal can overheat or dry out when stranded above the water. Where winters are severe, they can freeze during low tide. If we consider how living things exchange heat and moisture with their environment, we can understand how exposure to air sculpts some obvious patterns in where things live on the shore.

57 Low tide on a bright, warm day may mean easy tide-pooling for us, but it can spell disaster for marine plants and animals stranded on the rocks above the receding water. The sun-bleached algae in this picture cooked too long in the air and died *(opposite)*.

58 At high tide, water covers most of the plants and animals living on the shore. This picture shows the white foam on a wave rushing through the half-submerged sea cave in the middle of the large rock island.

59 The same scene shown in plate 58, viewed at low tide. The rock
"island" is now a peninsula, connected to the shore by an expanse of stony
terraces and boulders covered with sea life. The cave in the middle of the
rock now appears as a still, dark hole high above the water.

Heat and Moisture

Temperature limits life. Chemical reactions in living cells run more rapidly as the temperature goes up and slow down as it falls. Some reactions can be speeded up or slowed down more than others. In addition, biological molecules must be a particular shape to do their job but can jiggle into the wrong shape if they become too hot. Imagine the havoc wreaked on an assembly line in a factory if some of the workers decided to work much faster than their buddies, and others walked off the job! Temperature affects the biochemical assembly lines in living cells in a similar way. Furthermore, living things are mostly water, and water freezes at 0° Celsius (32° Fahrenheit), or at slightly lower temperatures if the water contains dissolved salts or other molecules. Although some plants and animals can survive the formation of ice crystals in their bodies, many cannot. Some species can tolerate a wide range of body temperatures, whereas others can only function within a narrow temperature span.

Because we humans regulate our body temperature, we can cope with both high and low environmental temperatures. In contrast, the animals and plants living on the shore heat up and cool down as the temperature of the environment around them rises and falls. What determines how hot or cold these living things become? Even though we are warm-blooded, our everyday experiences give us a feel for how bodies exchange heat with the world around them.

Heat and temperature are not the same thing. Molecules wiggle around more (they have more *kinetic energy*) when they are hot than when they are cold. *Heat* is the *total amount* of kinetic energy in a body of a given size or in a particular volume of fluid, whereas *temperature* is a measure of the *average intensity* of jiggling of the molecules in that solid or fluid. We can illustrate this difference between heat and temperature by considering chocolate milk. Think about the amount of cocoa powder that you add to a glass of milk as the heat you add to a body, and think of the color of the chocolate milk as the temperature. Set two glasses side by side and fill one to the brim with milk, but fill the other only halfway. Stir one spoonful of cocoa powder into each glass and compare their colors. Both batches of chocolate milk contain the same amount (one spoonful) of cocoa (analogous to heat), but the drink made with the smaller volume of milk is darker (analogous to having a higher temperature) than the drink made with the full glass of milk. In the same way, if a small creature takes up ten *joules* of heat, it reaches a higher temperature than would a larger creature that takes up ten joules of heat.[1] The temperatures of large beasts change more slowly than those of tiny creatures, just as a large pot of soup seems to take forever to heat up on the stove, whereas a single serving can be cooked quickly.

Intertidal plants and animals spend part of each day in water and part of it in air. It takes about 3,500 times more heat to raise the temperature of a cup of water by one degree than it does to increase the temperature of a cup of air by one degree, so the temperature of the ocean changes far more slowly than the temperature of the air above it. Wade into the sea on a chilly morning, and both the air on your face and the water around your toes are cold. Wade into the surf again a few hours later in the noontime heat, and your feet are once again in frigid water even though hot air surrounds your head. When the plants and animals on a shore are submerged, they soak in a bath of nearly constant temperature, but when they are in air, the temperature around them can change rapidly.

---

1. It takes 4.2 joules of heat to warm up one gram of water one degree Celsius, from a temperature of 14.5° C to a temperature of 15.5° C.

Living things exchange heat with the environment around them in a number of ways. One mechanism of heat exchange is radiation. Bodies radiate heat. The coils in your toaster radiate heat, which browns the bread. The sun radiates heat, some of which is absorbed by objects on Earth. You feel hotter when you stand exposed to the bright sun than when you hide from solar radiation in the shadow of a tree. In the same way, a snail stranded at low tide on the sunny side of a rock warms up more rapidly than a snail on the shady side of the rock. We can stay cooler if we wear white clothing, which reflects away some of the radiation from the sun, than if we wear black. The shiny, pale shells of some intertidal animals can have the same effect as my white shirt and hat. Bodies not only absorb heat radiated from other bodies, but they also lose heat by radiation. We feel much colder on a starry winter night than we do on a cloudy night at the same air temperature. That's because on a clear night we radiate heat away to the universe, whereas on an overcast evening, the clouds reflect back to us some of the heat that we, the things around us, and the Earth radiate. Likewise, plants and animals on the shore are in greater danger of freezing during nighttime low tides that occur under a black, clear sky than under a blanket of clouds.

Another way that living things lose or pick up heat from the surrounding environment is via conduction. Heat is conducted between bodies and the things they touch. When you sit in a bathtub of hot water, heat is conducted from the hot water into your cooler body. When you get out of the tub and step barefooted onto the cold bathroom floor, heat is conducted from your warm body into the colder tiles. The bigger the temperature difference between you and the floor, the faster your feet lose heat. Step onto a fluffy rug, and your feet don't feel as frigid because the air-filled pile of the rug conducts heat away from you more slowly than the tiles do. Some materials conduct heat rapidly, while others conduct heat slowly. Rock, water, and living tissues conduct heat more quickly than air. Therefore, a barnacle on the shore at low tide tends to be the temperature of the rock to which it is stuck rather than the temperature of the air around it.

The bigger the area of contact between a body and the things around it, the more rapidly heat is conducted between them. When you try to walk barefooted across the blistering asphalt of a parking lot roasting in the summer sun, much less heat is conducted into your feet if you leap and prance on tiptoe than if you tread flat-footed across the pavement. In the same way, heat is conducted more rapidly from a sun-baked rock into a limpet (plate 12) across its large, flat foot, plastered to the stone, than into a mussel hanging from the rock by a few skinny byssal threads (plate 37).

If the temperature of a plant or animal is greater than the water around it, a halo of water right next to its body warms up as heat is conducted into the water from the body. As that halo of water gets warmer, the temperature difference between the halo and the body becomes smaller, so the individual loses heat more slowly. However, if you replace that warmed water with new, chilly water, then the rate that the body loses heat increases again. In chapter 5 we explored how moving water carries dissolved substances to and from the living things on the shore. Moving water transports the heat that it holds in the same way, so if you reread chapter 5 using the word "heat" instead of "substance," you'll understand how currents and waves affect the temperatures of seashore plants and animals when they are submerged. For example, the faster the water flow in an individual's neighborhood and the thinner the boundary layer of slowed fluid around its body, the more rapidly the individual will return to the temperature of the ocean after it is submerged by the rising tide. Similarly, plants and animals surrounded by air exchange heat more rapidly with the air when the wind is blowing than when the air is still, but the effects are not as obvious as in water because air conducts heat so slowly.

Evaporation is a great way to cool off in air. When a gram of liquid water evaporates to water vapor at the surface of a damp body, it uses up about 2,500 joules of heat. We use evaporative cooling when we sweat, so we know from experience that the air conditions around us determine how effective evaporative cooling can be. For example, our sweat evaporates more quickly and we lose heat more rapidly if the air around us is dry than if it has a high moisture content. That is why we feel much hotter on a steamy, muggy afternoon in New Orleans than we do at the same temperature in the parched desert air of Tucson. In either city, when you are perspiring, you cool off more slowly if you sit around in still air than if you sit in front of a fan. That occurs because the layer of air right around your body becomes laden with water vapor as your sweat evaporates. Since you are surrounded by your own personal halo of high-humidity air, the sweat on your skin evaporates more slowly. When the wind blows or you fan yourself, you sweep away that boundary layer of hot, humid air and replace it with new air that is carrying less water vapor. In the same way, living things on the shore at low tide cool off by evaporation more rapidly if the wind is howling than if the air is still.

Unfortunately, when water evaporates from intertidal plants and animals, they dry out. Some species can survive being dehydrated; others cannot. Whether an individual is in danger of deadly drying depends on its structure and on where it sits on the shore. If you consider what to do with your wet socks after you have been wading in the ocean, you'll understand some of the factors that affect how rapidly intertidal plants and animals lose water. Water evaporates more quickly when the temperature is high, so if you lay your socks on a hot boulder in the sun, they dry faster than if you drop them on a cool surface in the shade. Likewise, a barnacle baking on the sunny side of a rock is in greater danger of drying up than a barnacle hiding in a cool shadow. If you hang your socks from a clothes line high up in the wind, they dry more quickly than if you leave them lying in the slowly moving air along the ground. In the same way, tall plants and animals that stick up into the breeze above their neighbors lose water more quickly than short individuals hugging the rock or hiding in the still air of crevices. When you wad your socks together into a ball, they stay soggy much longer than if you lay each one out by itself. Similarly, seaweeds piled up among their neighbors (plate 60) stay damp when isolated algae don't (plate 57). If you put your socks in a bag, they dry more slowly than if you let them sit out in the air. A mussel's moist tissues inside its shell dry out more slowly than a sea anemone's body, which has no such covering. If you lay one sock out on the surface of the beach but cover its mate with sand, the sandy sock stays wet longer. So do sea anemones coated in sand when the tide recedes (plate 73). Your big, thick, wooly socks dry more slowly than the tiny, thin socks of a toddler, just as massive kelp take longer to lose all their water than do diminutive sporelings.

With wet socks, frigid bathroom floors, and sweaty brows in mind, let's look at what happens to the algae and animals left high and dry on the shore as the tide recedes.

Whether Weather Matters

How perilous a period of exposure to air can be for intertidal plants and animals depends on the weather conditions when the tide is out. If they are stranded out of water on a hot, clear day when the air is dry and the sun is bright (plate 60), the danger of baking to death is far greater than on a gray, foggy day when the air is chilly and laden with moisture (plate 61). The risk of drying to a crisp is much worse if the wind is howling (plate 62) than if the air is still. Warm, sunny weather can be lethal if low tide occurs at high noon but not if the water recedes in the cool of evening.

The longer an intertidal animal or plant is exposed to dangerous terrestrial conditions, the greater its risk of drying out, cooking, or freezing. During part of each month, the shore is exposed to *spring tides,* when the low tides are lower and the high tides are higher than average. At another time of the month, *neap tides* occur, when the difference between high and low water is much smaller. During neap tides, creatures living high on the shore might not get wet, even when the tide is in, while those low on the shore may remain under water even when the tide is out. Spring tides can be perilous times: animals and plants in the middle of the intertidal zone are stranded out of the water far longer than during neap tides, while living things that reside lower on the rocks must face the rigors of exposure to the air only during spring tides. A low spring tide occurring at midday during hot, sunny weather can be a fatal formula for many intertidal plants and animals.

On a given day, while the tide is out, surf and weather conditions can affect the duration of risky terrestrial conditions. For example, on blustery days, as clouds blow overhead, they may obscure the sun for minutes

60 The sunny faces of rocks are festooned with photosynthesizing algae and seagrass. Plants in the middle of such dense aggregations are shielded from the wind and shaded from the sun by their neighbors, so they stay wetter and cooler than do isolated individuals *(opposite)*.

61 Emerald seagrasses and chocolate-colored algae exposed to chilly
wet air on a foggy day can stay moist and cool for hours *(opposite)*.

62 Grass sways in the wind whipping across the cliffs overlooking a
churning sea. Intertidal animals and plants exposed at low tide to rapid
air movement like this can dry out quickly *(above)*.

63 Shadows move across the shore as cloud after cloud is blown across the sky on a blustery winter day.

at a time and temporarily shade living things on the shore (plate 63). Furthermore, on days when the ocean swells hitting the shore are large, the spray from waves crashing onto the rocks splatters animals and plants above the tide with cool water (plate 64). Whether these brief showers, which bring relief from heat and dehydration, occur every few seconds, every few minutes, or less often, depends not only on the size of the waves that day, but also on where an individual sits on the shore (see chapter 3).

Spatial Patterns of Microclimates on the Shore

One obvious pattern in the distribution of plants and animals on the shore is *vertical zonation* (see plate 13). High on the rocks we find a zone dominated by acorn barnacles and periwinkle snails. Just below the barnacles, an obvious horizontal band of mussels and goose barnacles runs along the shore. Lower still, space on the rock is dominated by turfy algae, and below that zone we find larger seaweeds and surfgrasses as well as starfish and sea anemones. The complex interplay of physical factors and biological interactions that produce this zonation is the subject of other books (see the "Additional Reading" section). The point to ponder here is that the higher a creature lives on the shore, the more time it spends out of the water when the tide is low. Barnacles and periwinkle snails, which live highest on the rocks, can retreat into their shells when the tide goes down. They can also physiologically withstand the heat they experience when stranded for hours on sun-baked rocks. In contrast, a sponge or starfish would dry up and die quickly if placed high above the water in the barnacle zone but can survive the brief exposures to air it experiences in its habitat low down on the shore. These different abilities to survive in air affect the biological interactions among different shoreline species. For example, mussels can outcompete many other types of creatures for space on the rock, but mussels can't survive as high above the water as barnacles can. Therefore, barnacles have a mussel-free refuge high on the shore where they can persist. Mussels are the favorite food of the large colorful starfish (plate 43) that huddle below them on the shore. Because these starfish cannot tolerate being out of the water as long as mussels can, there is a zone on the shore above the reach of the starfish where dense beds of mussels can survive.

When waves crash onto a stony shore, water splashes up the seaward faces of the rocks, bathing the plants and animals clinging to those surfaces (plate 64). Such periodic showers of water permit some marine species to survive in positions above the high-tide mark, so the zones of intertidal animals and plants are wider and higher on the ocean-facing sides of rocks than on the unsplashed shoreward sides (plate 65). For the same reason, the more wave-exposed the site, the higher the spray and the wider the zones of intertidal life.

Pools of seawater trapped in depressions in the rock when the tide recedes provide aquatic habitats in the intertidal (plate 66). Delicate marine life that would perish if exposed to air can survive in these tide pools (plate 67). The higher a pool is on the shore, the less frequently it is flushed out by the rising tide. Stagnant pools high on the shore can be dangerous places that might overheat or freeze, become too salty as they dry up or too dilute as they fill with rainwater, or turn foul if polluted by bird droppings. The animals in a tide pool can sometimes suffocate if they use up all the oxygen in the water, but beasts in pools containing seaweeds are less likely to suffer this fate when the sun is shining because oxygen is produced during photosynthesis.

Cracks and crevices in the rock can provide cool, damp refuges for creatures left behind by the tide (plate 68). Animals can also hide from the sun and wind in the spaces within mussel beds (plate 6) or under the forests of seaweeds that blanket the rocks (plate 60). The misty, chilly

64 When a wave slams into a rock, water sprays high up on the seaward
face of the rock.

65 The zones dominated by different types of animals and plants (described on page 17) can be seen on the large rock in the foreground at low tide. Notice that the band of black mussels is much wider and extends up higher on the right side of the rock than on the left. When the tide is in, waves first hit the right side of this rock, sending a shower of spray up that seaward-facing surface. This splashing allows mussels and barnacles to survive much higher up, and thus the zones they dominate are wider on the seaward sides of rocks (opposite).

66 Pools of seawater are trapped in depressions in the rock as the tide recedes (above).

67 Tide pools provide aquatic refuges for creatures that cannot tolerate exposure to air, like the fragile striped worm and the feathery hydroids shown here. Hydroids are related to sea anemones. The golden brown featherlike structures in this photograph are the external skeletons of colonies made up of many tiny tentacled hydroid animals that are connected to each other and share a common gut. Hydroid colonies come in different forms: the golden threads in plate 34 are colonies of a different species *(above)*.

68 Chalky limpets and black snails crowd into a shady, moist crevice at low tide, alongside a few sand-encrusted sea anemones. Notice the battered shell of the limpet near the bottom of the picture. The dark dents are places where chips were broken out of the shell, probably by pebbles thrown against the shore by waves *(opposite)*.

recesses of a sea cave (plate 69) harbor animals and algae that must remain moist to survive (plate 16).

Seaweeds and seagrasses need sunlight for photosynthesis, as do certain animals, such as some species of sea anemones (plate 50), that have symbiotic single-celled algae living in their bodies. These algae carry out photosynthesis, producing energy-rich sugars used by both the algae and the host animal. For the plants and animals of the shore that use sunlight for photosynthesis, there is a trade-off between basking in light versus hiding from the sun to minimize the danger of drying out or roasting. The sunny sides and tops of rocks are often covered with turfy species of algae that hang on to water like a wet mop and that can also survive being dehydrated. Some seaweeds hold pools of water in their baggy bodies (plate 70). In contrast, the species of algae hiding in the shade under ledges and in caves (see plates 7 and 14) can photosynthesize in low light but can't tolerate being dried or cooked. Even out in the bright sun, within dense clumps of algae, light can be reduced so much that photosynthesis is limited (plate 71). Sea anemones can benefit from the photosynthesis of their symbiotic algae (plate 8) but can also survive in dark places without these symbionts (plate 72) because sea anemones are carnivores that can make a living by capturing prey.

Sunburn (damage to living things caused by ultraviolet radiation from the sun) can be a problem for intertidal plants and animals. For example, ultraviolet light can kill algal spores and germinating young seaweeds, especially if they are cold. An obvious way to avoid sunburn, both for us and for intertidal creatures, is to hide in the shade. In addition, various intertidal animals, like sea anemones and sea urchins, have sunscreen molecules in their body tissues that protect them from damage by ultraviolet radiation. The spawned eggs and the tiny embryos and larvae of marine animals that

69 An opening in the rock leads into the dark, damp recesses of a cave that provides shelter from sun and wind *(opposite)*.

70 Reservoirs of water can be seen through the translucent walls of saclike algae.

71 Dense stands of sea palm algae absorb sunlight. Only about 10 percent
of the light hitting this forest of fronds makes it through the canopy into the
dark spaces between the stipes.

72 This ghostly white sea anemone lurks in the dark in the deepest bowels of the cave in plate 69. It is a member of the same species as the green anemones in plates 8 and 50 but has shed its photosynthetic symbiotic algae and lost its own pigments.

are washing around in the ocean are also vulnerable to damage by ultraviolet radiation if they are up near the water's surface, and many of them also protect themselves with sunblock compounds. While some types of plants and animals can make their own sunscreens, some animals accumulate sunblock compounds from the seaweed they eat.

What to Do When the Tide Goes Out

Many intertidal animals exhibit behaviors that help them survive exposure to air. Mussels and barnacles close their shells, and sea anemones retract their tentacles down into their bodies. Some anemones cover themselves with a coat of sand and bits of shell, held in place on special wartlike, sticky bumps in rows along their bodies (plate 73). Crabs retreat into shady crevices or pools of water, while snails aggregate, bunching together in moist places (plate 74). Animals that lay eggs on the substratum deposit them in dense masses hidden in damp, cool places where they are protected from drying out and from damage by ultraviolet radiation (plate 75).

A Patchwork of Microclimates

When the complex terrain of a rocky coast is submerged at high tide, the water-flow environments experienced by various plants and animals on the shore can be drastically different from each other, as explained in chapters 3 and 4. Similarly, when the complex terrain of a rocky coast is exposed to the air at low tide, a rich tapestry of diverse microclimates develops (plate 76). The cool, wet conditions experienced by living things in pools, caves, and crevices can be nothing like the dangerous heat, wind, or ice faced by plants and animals on the tops of boulders.

73 Densely packed crowds of pink-tentacled sea anemones can survive higher on the shore than their solitary big green relatives shown in plate 8. At low tide, the sea anemones retract their slender, thin-walled tentacles that are so vulnerable to water loss. These animals have plastered their body surfaces with a coating of sand and bits of shell that protects them from drying out rapidly *(above)*.

74 Snails, which disperse across the rocks to graze on algal scum when they are submerged, gather together in groups in damp refuges when the tide recedes. Wind-tunnel experiments have shown that snails and limpets in aggregations like this lose water more slowly than do solitary individuals sitting alone on a bare rock *(opposite)*.

75 Some intertidal species glue clumps of eggs to the bottom rather than release them into the water. Eggs packaged into a dense mass, like the pink ones pictured here, dry out more slowly than would isolated eggs when exposed to the air *(above)*.

76 In the convoluted terrain of a rocky shore, a hodge-podge of many different microclimates develops during low tide *(opposite)*.

7

CHANGES

Physical Change

The rocky-shore environment changes with time. Waves swash back and forth every few seconds, exposing animals and plants on the rocks to transient pulses of force. Clouds blow across the sky and hide the sun every few minutes, providing brief respites from the heat. In the span of hours, tides rise and fall, the fog burns off and rolls back in, and day fades into night. Calm mornings give way to blustery afternoons. The weather changes from one day to the next as storms hit the coast and then move on.

Over the course of a month, the tides wax and wane. During neap low tides, only a small strip of the life at the edge of the sea is exposed to air. A couple of weeks later, dramatic spring tides expose a wide swath of the shore to the dangers of sun and breeze.

Conditions vary as the seasons come and go. Wind and storms across the ocean generate huge waves that batter the coast in winter, whereas small swells on the summer sea wash lazily across the shore. Plants and animals that experience long periods of daylight in the heat or fog of

78 Winter surf pounds the coast.

summer are later subjected to the rainy and icy days of winter, with only a few hours of light. Global patterns in ocean currents and weather, which change from year to year and decade to decade, also affect the local environment of our small stretch of rocky coastline. Some years are especially harsh and stormy, whereas others are unusually mild. Some years are warmer than average, and some are cooler. Wet years can follow years of drought.

These examples illustrate that physical changes on the shore happen on many time scales, from seconds to decades.

Biological Change

The living neighborhood of a plant or animal on the shore also changes with time. Grazers and predators come and go, while living things attached to the rock grow and jockey for space (plates 79–81). Plants and animals age, accumulating and repairing damage and producing offspring. Larvae settle in the neighborhood, metamorphose into their bottom-dwelling forms, and take up residence in the community (plate 82). A patch of bare rock scraped clean by a wave-tossed log (plate 48) is filled in by a succession of species, the early pioneers eventually being overgrown by more competitive colonists.

Lush gardens of seaweeds that thrive in the summer are ripped from the rocks and washed ashore by winter waves. Beaches that were bare and clean can be buried in mats of dead algae (compare plate 2 with plate 83). The knots of beach wrack strewn on the shore (plate 84) are invaded by small crustaceans and insects that help break down the smelly, rotting algae in a matter of weeks.

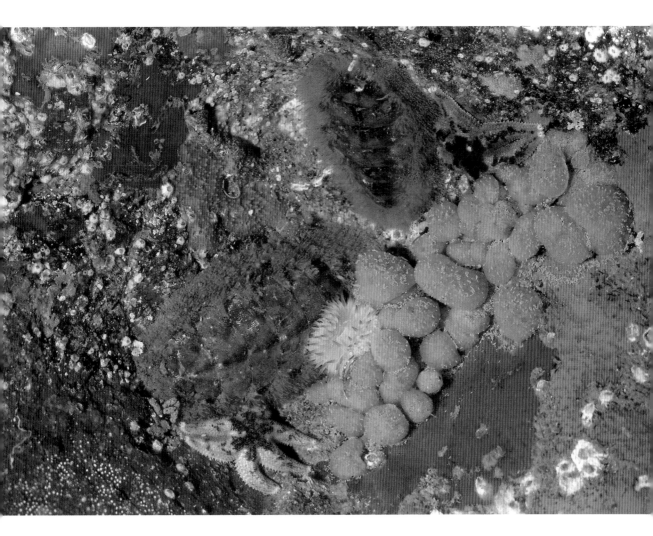

79 A small tide pool in the cave is lined by a carpet of bright red sponges, but just a year earlier it was dominated by different species. The next two plates chronicle the biological changes in this little pool during the months preceding this photograph (*opposite*).

80 A year before the photograph in plate 79 was taken, the tide pool in the cave harbored a diversity of inhabitants, including bulbous beige sea squirts, a tiny tan starfish, a pink-tentacled sea anemone, a multitude of barnacles, and a few small crimson sponges (*above*).

81 Six months after the picture in plate 80 was taken, two large predatory starfish are visiting the pool. Many barnacles have perished, and the area of the pool floor covered by sea squirts is much smaller, while the sponges have expanded greatly, annexing more territory *(above)*.

82 Diminutive juvenile goose barnacles can be seen among the larger, older animals. One important cue that stimulates many types of larvae to settle in suitable habitats is the presence of members of their species already thriving in particular spots on the shore *(opposite)*.

83 By the time winter is over, much of the sand on the beach has been washed away, and heaps of broken seaweeds pile up at the edge of the sea. In the calm of summer, the sand will be deposited back onto the beach.

84 *Beach wrack,* a tangled mat of broken algae, lies drying and rotting on the beach after winter storms. The long, hoselike stipes of subtidal bull kelp are knotted around bright green seagrass blades and purple algal fronds ripped from the intertidal rocks.

163

Living Things Respond to Change

Plants and animals living on a wave-swept shore respond to the ever-changing conditions they encounter. Flexible seaweeds give like shock absorbers with every wave that hits them, while crabs scurry into crevices to avoid high surf (plate 39). Sea anemones flatten themselves into pancakes hugging the rock when they feel rapidly flowing water, and they retract their delicate tentacles when stranded in air at low tide (plates 32 and 73). Mussels and barnacles close their shells when the tide goes out, slowing the rate at which they lose water (compare the mussels and goose barnacles with their shells pulled shut in plate 6 with the more vulnerable animals that are open and feeding in plates 52 and 54). When the water recedes at low tide, snails and limpets aggregate in cool cracks, clamping their shells down tightly onto the rock to seal their moist bodies away from the air (the wet, fleshy foot of the crawling snail in plate 44 is safely tucked away from the sun and wind in plate 68).

Over longer periods—days to months—some plants and animals can grow or remodel themselves in response to environmental conditions. Mussels that survive the first storms of winter secrete additional byssal threads (plate 37) to anchor themselves more securely to the rock. Some algae, if damaged, repair themselves with new tissues that are tougher than their old tissues.

Many species of seaweeds grow flat straplike blades in habitats exposed to rapidly moving water but produce wide, ruffled blades at calmer sites. The straplike blades of the plants in exposed habitats are pushed together into slim, streamlined bundles by moving water and thus produce small wakes and experience low drag. This straplike form enhances the survival of seaweeds that experience fast flow. In contrast, the curly blades of the algae at calm sites do not fold together into slender bundles in water currents but remain spread out in the sunlight. These ruffly blades flutter in the flow, stirring things up and enhancing their uptake of dissolved materials from sluggish currents. However, a big wake forms behind a wide clump of curly blades spread out and flailing in the flow. Seaweeds with the calm-water curly form would experience dangerously high drag if they were living in fast flow. Ruffly seaweeds transplanted from habitats with sluggish flow into current-swept places become straplike (if they don't wash away), and straplike seaweeds transplanted from exposed sites to calm habitats become ruffly. The mechanical forces on a growing algal blade affect the shape it assumes, sculpting it into a form well suited to its flow environment.

Pacing and Timing

The diverse animals and plants living side by side on a stretch of shoreline differ in their *life-history traits:* that is, they differ in the rates at which they grow, the timing of their reproduction, how they allocate their resources between growing and reproducing, and their modes of dispersal to new sites. The timing of these life events relative to the pace of changes on the shore often correlates with the mechanical architecture of a living thing. Strong, tough animals and plants that invest in making unbreakable structures, such as protective shells or woody stipes, often grow slowly and live for a number of years, reproducing more than once in their lifetime. In contrast, some soft, fleshy seaweeds grow and mature very rapidly but are easily ripped from the rocks. Even though their floppy bodies seem badly engineered for life on a wave-swept shore, such species can be quite successful there if they grow and reproduce quickly in the calm of

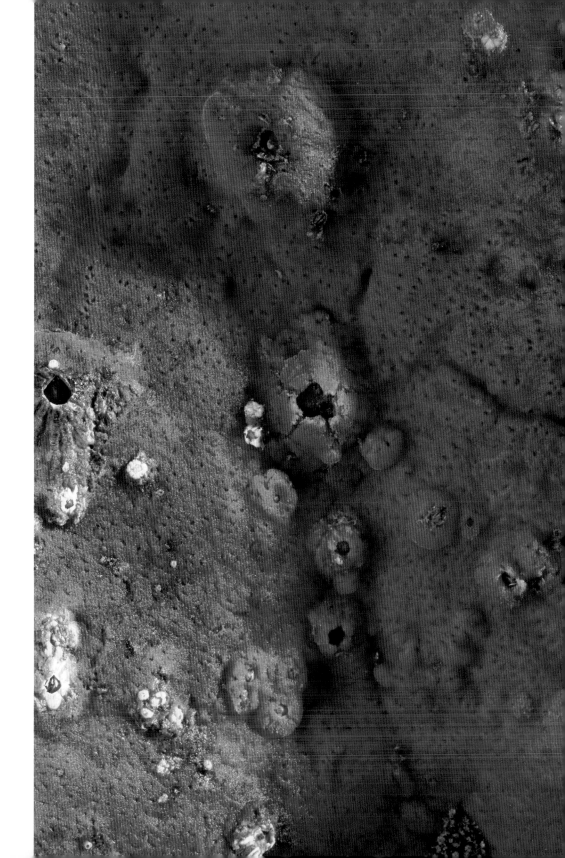

summer, before they are swept away by the predictable storms of winter. Of course, this weakling strategy only works at wave-exposed sites if the spores, larvae, or rafting bits of adults can recolonize the shore, grow, and reproduce during the next calm season.

A wave-swept shore is a dynamic place. The physical world and the fabric of life on the shore are in constant flux (plate 85), changing by the second, changing by the decade.

85 When animals die, their abandoned shells provide surfaces where other creatures can live. Here a red sponge grows over the husks of dead barnacles *(opposite)*.

8

STEPPING BACK

The Richness of a Place

Step back and contemplate this site where the ocean meets the land. Our understanding of a place like this is easily biased by the spatial scale at which we humans experience it. However, if we approach the natural world around us armed with a handful of physical principles and a willingness to imagine the experiences of creatures of different sizes, we can recognize the rich tapestry of habitats woven into a single site. We can appreciate the specific challenges to survival in each spot and understand many of the mechanisms that enable creatures to persist in the face of those challenges.

86 As we leave the sea and climb back up the hill, returning to our familiar terrestrial world, we pass through clumps of brush and flowers. We now can understand how these land plants interact with the wind if we apply to them the insights we have gained by studying how water moves around kelp. Likewise, the principles of structural design that let us figure out why some intertidal plants and animals are deformable or breakable, while others are not, enable us to analyze how living things on land respond to mechanical forces. What we've learned about how bodies heat up and dry out applies to the plants and animals that live in terrestrial habitats as well as to the inhabitants of the intertidal zone *(opposite)*.

So What?

By investigating how water moves through complex marine habitats and how animals and plants of different designs withstand and use that flowing fluid, we learn principles that can help us understand how other types of plants and animals function in different sorts of habitats. Furthermore, if we understand the mechanisms by which weather and waves affect the survival of living things, we can better predict the biological consequences of global climate change, which is leading to more storms, bigger waves, and hotter days.

Can any *practical* good come from studying how seashore animals and plants interact with their physical surroundings? Even for people who don't care about the environment, there are compelling reasons to study the physics of marine creatures. For example, man-made structures deployed in the ocean, like docks or offshore drilling rigs, are soon colonized by sea life. Our measurements of the hydrodynamic forces on various types of marine plants and animals make it possible to calculate the additional loads that must be sustained by man-made structures overgrown by such creatures and hit by waves and currents. Designing marine structures to sustain the added load from "biofouling" can avert structural failure.

Biomimetics is a blossoming field of engineering in which the design of man-made materials, devices, and structures is inspired by the design of living things. Engineers do not build exact copies of biological structures but rather study the principles that govern how they work. We then can use those principles to design something new. The microscopic architecture of sea shells inspires the design of tough new composite materials. The chemistry of barnacle glue and the mechanical behavior of mussel byssal threads teach us how to stick things together. Studying the mechanics of crabs running underwater and in air can help us build stable, dynamic amphibious robots that can navigate in waves for tasks such as clearing a beach of mines.

A Personal "So What?"

Vistas of a rocky coast can be stunning, but a deeper understanding of how the shore functions makes such a place all the more beautiful for those who take the time to study it. As the nineteenth-century British philosopher Herbert Spencer said, "Those who have never entered upon scientific pursuits know not a tithe of the poetry by which they are surrounded."[1]

1. Herbert Spencer, *Education: Intellectual, Moral and Physical* (New York: D. Appleton, 1890), 83.

Additional Reading

User-Friendly Books about Related Topics

Books about the Ecology of Rocky Shores

Carefoot, Thomas. *Pacific Seashores: A Guide to Intertidal Ecology.* Seattle: University of Washington Press, 1977.

Rosenfeld, Anne W., with Robert T. Paine. *Intertidal Wilderness: A Photographic Journey through Pacific Coast Tidepools, Revised Edition.* Berkeley: University of California Press, 2002.

Natural History Guides to the Pacific Seashore of North America

Abbott, Isabella A., and George J. Hollenberg. *Marine Algae of California.* Stanford, CA: Stanford University Press, 1976.

Kozloff, Eugene N. *Seashore Life of the Northern Pacific Coast.* Seattle: University of Washington Press, 1983.

Morris, Robert H., Donald P. Abbott, and Eugene C. Hatterlie. *Intertidal Invertebrates of California.* Stanford, CA: Stanford University Press, 1980.

Ricketts, Edward F., Jack Calvin, and Joel W. Hedgpeth. *Between Pacific Tides,* 5th ed. Revised by David W. Phillips. Stanford, CA: Stanford University Press, 1985.

Publications about How Structures Work and Fluids Flow

Gordon, J. E. *Structures: Or Why Things Don't Fall Down.* New York: Da Capo Press, 1978.

Koehl, M. A. R. "The Interaction of Moving Water and Sessile Organisms." *Scientific American* 247 (1982): 124–32.

Shapiro, Ascher H. *Shape and Flow.* New York: Anchor Books, 1961.

Vogel, Steven. *Life's Devices.* Princeton, NJ: Princeton University Press, 1988.

A Book with Historical Perspective

Gladfelter, Elizabeth Higgins. *Agassiz's Legacy: Scientists' Reflections on the Value of Field Experience.* New York: Oxford University Press, 2002.

Books for Those Who Want to Delve into These Topics More Seriously

Books about the Physics of Organisms and Habitats

Denny, Mark W. *Biology and the Mechanics of the Wave-Swept Environment.* Princeton, NJ: Princeton University Press, 1988.

Monteith, J. L., and M. H. Unsworth. *Principles of Environmental Physics,* 2nd ed. London: Edward Arnold, 1990.

Vogel, Steven. *Comparative Biomechanics: Life's Physical World.* Princeton, NJ: Princeton University Press, 2003.

Books about Marine Biology and Marine Ecology

Jumars, Peter A. *Concepts in Biological Oceanography.* New York: Oxford University Press, 1993.

Levinton, Jeffrey S. *Marine Biology: Function, Biodiversity, Ecology.* San Francisco: Benjamin Cummings, 2001.

Nybakken, James W. *Marine Biology: An Ecological Approach,* 5th ed. New York: Harper, 2001.

Paine, Robert T. *Marine Rocky Shores and Community Ecology: An Experimentalist's Perspective.* Oldendorf/Luhe, Germany: Ecology Institute, 1994.

Index

Page numbers in italics refer to photographs, diagrams, and their captions.

Design and composition   Victoria Kuskowski

Diagrams   Dartmouth Publishing, Inc.

Index   Andrew Joron

Text   11/16.5 Granjon

Display   Walbaum

Printer and binder   C&C Offset Printing Co.